T0064665

CHEMISTRY

QUESTIONS and ANSWERS

Amin Elsersawi, Ph.D.

authorHOUSE®

AuthorHouse™ LLC
1663 Liberty Drive
Bloomington, IN 47403
www.authorhouse.com
Phone: 1-800-839-8640

© 2014 Amin Elsersawi. All rights reserved.

No part of this book may be reproduced, stored in a retrieval system, or transmitted by any means without the written permission of the author.

Published by AuthorHouse   07/14/2014

ISBN: 978-1-4969-2704-0 (sc)
ISBN: 978-1-4969-2716-3 (e)

Any people depicted in stock imagery provided by Thinkstock are models, and such images are being used for illustrative purposes only. Certain stock imagery © Thinkstock.

This book is printed on acid-free paper.

Because of the dynamic nature of the Internet, any web addresses or links contained in this book may have changed since publication and may no longer be valid. The views expressed in this work are solely those of the author and do not necessarily reflect the views of the publisher, and the publisher hereby disclaims any responsibility for them.

Introduction

This book helps students and readers visualize the three-dimensional atomic and molecular structures that are the basis of chemical action. An integral part of the text is to develop an explanation to hybridization which introduced to explain molecular structure when the valence bond theory failed to correctly envisage them.

Dr. Elsersawi presents the quantum theory of the electronic structure of atoms and focuses on the electronic structures and reactivity of atoms and molecules. Many questions and answers of chemical components are introduced, using molecular orbital, and hybridization of orbitals.
This text maintains strong exercises, scientific integrity and consistency in presentation. Atoms, molecules, ions, chemical formulas and equations, chemical bondings, intermolecular forces, energies, electronegativity are offered to readers in effective and proven features – clarity of writing and explanation.

Overall, this volume constitutes an outstanding contribution for anyone with interest in chemistry. It is a useful reference for health professionals, practicing physicists, chemists, and materials scientists.

Contents

What is an atom?

The atom is the smallest unit of an element that holds the chemical and physical properties of that element. An atom has an electron cloud consisting of negatively charged electrons orbiting a dense nucleus. The nucleus contains positively charged proton and electrically neutral neutrons, Figure (1).

Figure (1): The atom

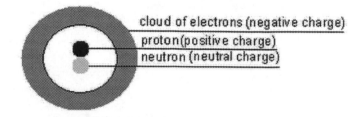

cloud of electrons (negative charge)
proton (positive charge)
neutron (neutral charge)

What is a molecule?

Molecule can found as a unit of two or more atoms held together by covalent bonds or ionic bonds. We shall talk about both types of bonds, but let talk first about binding of atoms. Atom is like a ball with three dimensions. So the attraction and repulsion between atoms are accomplished also in multidimensional manner. Let us consider the molecule of the ethylene Gas   C2H4. It has the molecular structure of H2C=CH2 and it can be shown in a Planner form of Figure (2):

Figure (2): Planner chemical form of ethylene molecule

What is a covalent bond?

A covalent bond is a form of chemical bonding that is distinguished by the sharing of pairs of electrons between atoms. The stability of covalent bonds is based on the attraction-to-repulsion forces between atoms. Let's take the following examples of methane, nitrate, and water, Figure (3).

Figure (3): Covalent bond of methane, nitrate, and water

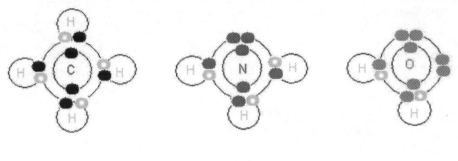

○ electron from hydrogen

What is an ionic bond?

An ionic bond is an electrical attraction between two oppositely charged atoms or groups of atoms. All atoms in molecules are trying to have their outer orbit in a stable condition or neutrality, i.e. 8 electrons. In order to gain stability they will either lose one or more of its outermost electrons thus becoming a positive ion (cation) or they will gain one or more electrons thus becoming a negative ion (anion). Thus atoms of negative charges attract with those of positive charges. That electrical attraction between two oppositely charged ions is referred to as an ionic bond. Most salts are ionic. Any metal will combine chemically with any non-metal to form ionic bonds that hold the molecule together.

Let's see the ionic bond in the reaction of sodium with chlorine.

Chlorine (on the right) pulls one valence electron from sodium atom, Figure (4) below.

Figure (4): Ionic bond of sodium chloride

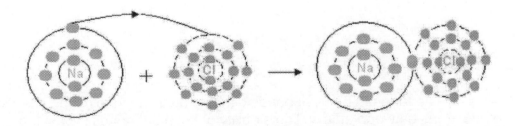

Noting that the valance in the chlorine is 7 (high negative), and the valance in the sodium is 1 (low negative), and thus the chlorine pulls the one electron from sodium.

What is the Valence Shell Electron Pair Repulsion (VSEPR) theory?

Valence Shell Electron Pair Repulsion (VSEPR) theory allows fairly accurate prediction of the 3-dimensional shape of molecules from knowledge of their Lewis Dot structure. In VSEPR theory, the position of bound atoms (ligands) and electron pairs are described relative to a central atom. Once the ligands and lone pair electrons are positioned, the resulting geometrical shape presented by the atoms only (ignoring lone pairs) is used to describe the molecule.

The electron density distribution for a molecule is best illustrated by means of a contour map. The contour map of the charge distribution for the lowest or most stable state of the hydrogen molecule is show in Figure (5) which shows a cross section of one molecule of the hydrogen.

Figure (5): Charge distribution of hydrogen molecule

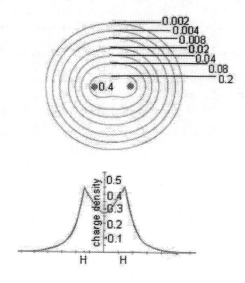

Another example of helium molecule is shown in Figure (6).

Figure (6): Repulsion between the two atoms of the helium molecular

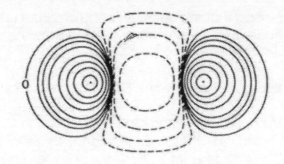

Valence Shell Electron Pair Repulsion (VSEPR) theory allows us to predict the 3-dimensional shape of molecules from knowledge of their Lewis Dot structure (it will be explained). In VSEPR theory, the position of bound pairs and lone pairs are described relative to the central of the atom. Once the bound and lone pair electrons are positioned, the resulting geometrical (physical) shape presented by the atoms only is used to describe the molecule.

Explain Lewis structures of atoms

The chemical symbol for the atom is surrounded by a number of dots corresponding to the number of valence electrons. Lewis Structures for ions of elements are shown in the table below.

Number of valence electrons	1		2		3	4	5	6	7	8
group	hydrogen	group I alkali metals	helium	group II alkali earth metals	group III	group IV	group V	group VI	group VII	group VIII
Lewis structure electron diagram	(H)	(Li)	(He)	(Be)	(B)	(C)	(N)	(O)	(F)	(Ne)

Lewis ionic structures of atoms

charge on ion	1+		2+	3+	4+	4-	3-	2-	1-	
electrons lost or gained	1e lost		2e lost	3e lost	4e lost	4e gained	3e gained	2e gained	1e gained	
group lost or gained	H+	group 1 1+ alkali metals	group II 2+ alkali earth metals	group III 3+	group IV 4+	group IV- 4-	group V 3-	group VI 2-	group VII 1- halogen	H- hybride
Lewis structure electron dot diagram	H^+	Li^+	Be^{2+}	B^{3+}	C^{4+}	C^{4-}	N^{3-}	O^{2-}	Fl^{1-}	

Let us take the molecule of beryllium chloride, BeCl2. The electronegativity difference between beryllium and chlorine isn't enough to allow the formation of ions as there are 4 electrons forming 2 pairs balancing each other.

Beryllium has 2 outer electrons because it is in group 2. It forms bonds to two chlorines, each of which adds another electron to the outer level of the beryllium.

Figure (7): Lewis structures

CH_4O

$[C_2H_2O_2]^+$

CH_5N

$[C_2H_3O_2]^-$

C_2H_6O

$[CH_6N_3]^+$

C_3H_6

$[CH_6N_3]^+$

$[C_3H_6N]^+$

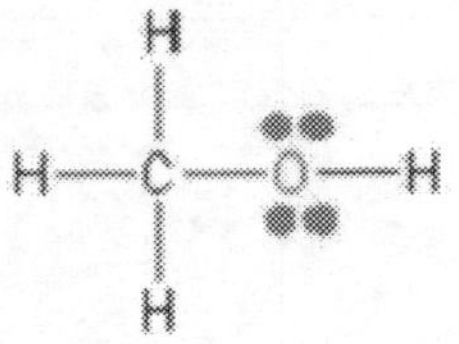

How many types of chemical bonds?

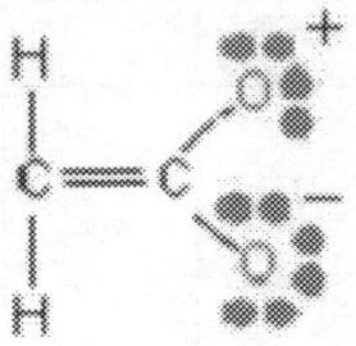

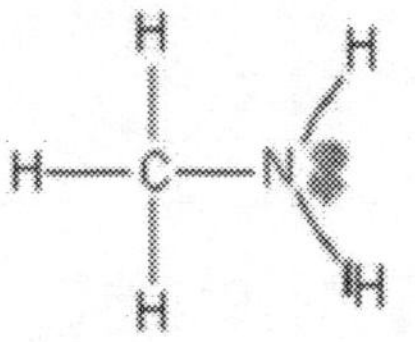

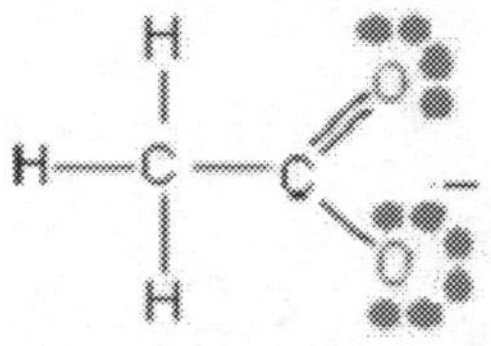

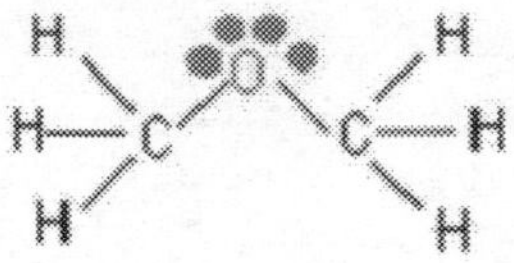

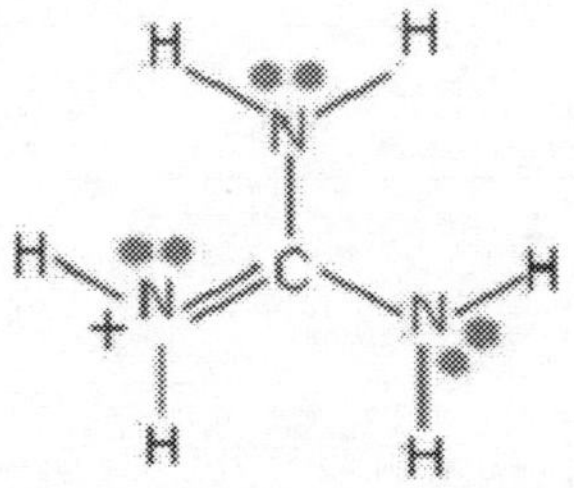

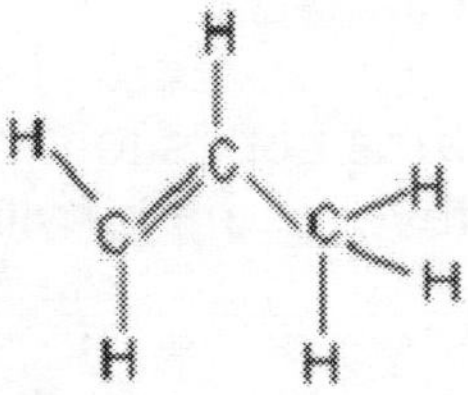

There are three types of chemical bonds are Ionic bonds, Covalent bonds, and Polar covalent bonds. Chemists also recognize hydrogen bonds as a fourth form of chemical bond, though their properties align closely with the other types of bonds.

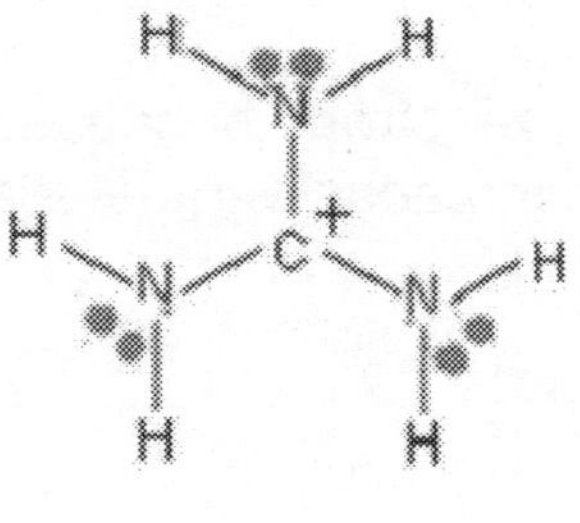

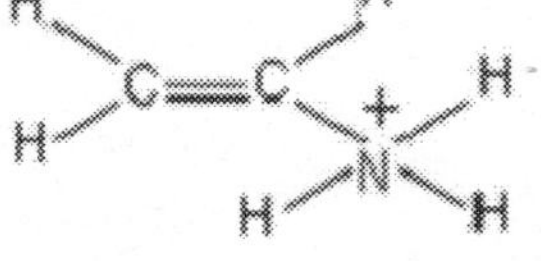

Ionic bonding is a type of electrostatic interaction between atoms which have a large electronegativity difference. There is no precise value that distinguishes ionic from covalent bonding, but a difference of electronegativity of over 1.7 is likely to be ionic, and a difference of less than 1.7 is likely to be covalent. An ionic bond occurs when one atom gains a valence electron from a different atom, forming a negative ion (anion) and a positive ion (cation), respectively. These oppositely charged ions are attracted to each other, forming an ionic bond.

A covalent bond results when two atoms "share" valence electrons between them.

A polar covalent bond is a covalent bond between two atoms where the electrons forming the bond are unequally distributed. This causes the molecule to have a slight electrical dipole moment where one end is slightly positive and the other is slightly negative. Water is a polar bonded molecule.

Can you show some bonds of different shapes?

This Molecule of boron trifluoride (BF_3) has 3 electrons in its outside orbit. They are pulled by the chlorine atom which has 7 electrons in its outside orbit. Note that boron in group 3 (13) and chlorine in group 7 (17) of periodic table. The outcome of attraction and repulsion is 3 bonding pairs which are at 120° to each other as shown in the corresponding plan.

This methane molecule (CH_4) has four bonding pairs arranged in a tetrahedral global arrangement. The angle between two hydrogen atoms is 109.5° calculated based on the central gravity of all of atoms that looks like regular three triangles with a based pyramid.

This ammonia molecule (NH3) has the nitrogen in the centre, and the angle between two hydrogen atoms is 107.8° as shown. The molecule is arranged in two lone pairs and one single electron. One of the two pairs is bonded to the two hydrogen atoms. The ammonia molecule has one lone pair and one bonding pair. The bonding pair is longer than the lone pair. Therefore, there is more repulsion between them than the two bonding pairs.

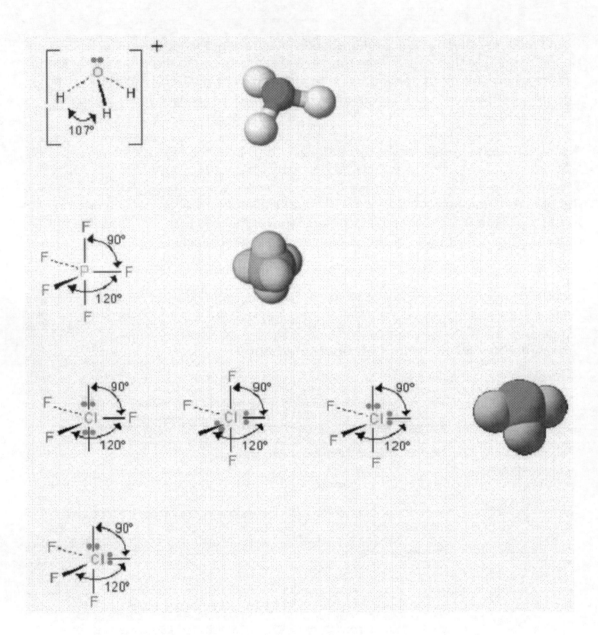

This ammonium ion (NH$_4^+$) has one hydrogen atom more than the normal capacity of three electrons in the nitrogen atom far most orbit. If the number of hydrogen atoms attached to the nitrogen atom is 3, the ammonium atom is neutral. The molecule is now positive as it carries one more hydrogen atom (proton). The ammonium molecule gains one electron from the orbit of the hydrogen atom, and in the same time, it gets one positive proton. Because the proton positivity is much larger than the negativity of the electron, therefore the ammonium molecule is positive.

What are the components of the cellular life?

1-   Water

- It comprises 60 - 90% of most living organisms (and cells).
- It is important because it serves as an excellent solvent and involves into most metabolic reactions.

2-   Carbohydrates

As the name implies, a carbohydrate is a molecule whose molecular formula can be expressed in terms of just carbon, oxygen and hydrogen and takes the formula of $C_xH_yO_z$, and if one hydrogen atom is lost the sugar becomes an aldehyde (in case of glucose) and is termed an aldose, or it is a ketone (in case of fructose) and is termed a ketose (combination of glucose and fructose is sugar, and will be discussed later). Monosaccharide contains either a ketone or aldehyde functional group, and hydroxyl groups on most or all of the non-carbonyl carbon atoms. For example, glucose has the formula $C_6H_{12}O_6$ and sucrose (table sugar) has the formula $C_{12}H_{22}O_{11}$. More complex carbohydrates such as starch and cellulose are polymers of glucose. Their formulas can be expressed as $C_n(H_2O)_{n-1}$ or $C_n(water)_{n-1}$.

2-1 Types of carbohydrates:

- Monosaccharide (e.g., glucose or fructose) - most contain 5 or 6 carbon atoms
- Disaccharide consists of 2 monosaccharide linked together.

Examples include sucrose (a common plant disaccharide is composed of the monosaccharide glucose and fructose), see Figure (8).

Figure (8): Glucose and fructose together (sugar), each of them has $C_6H_{12}O_6$ formula.

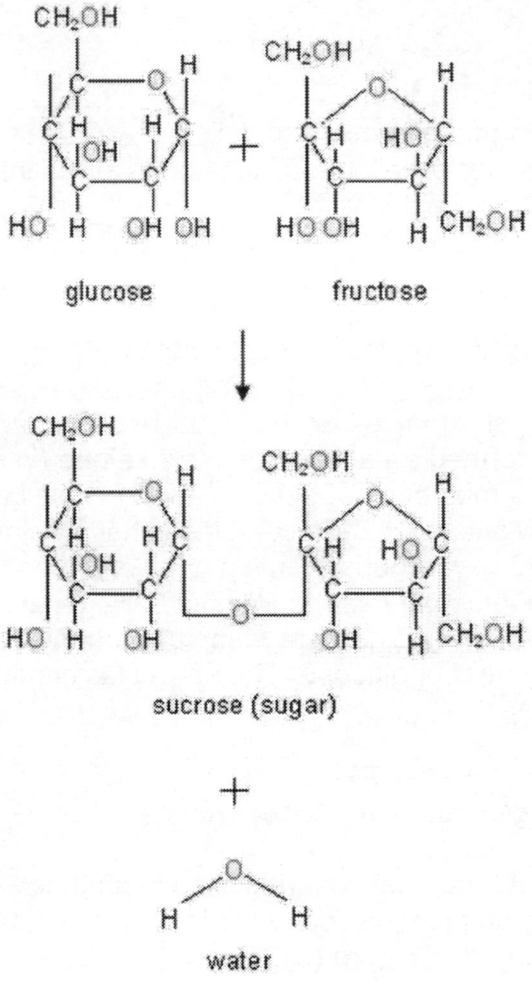

glucose   fructose

sucrose (sugar)

+

water

In linear form, two monosaccharide (glucose and fructose) linked together to form sucrose, Figure (9).

Figure (9): Glucose and fructose in linear form

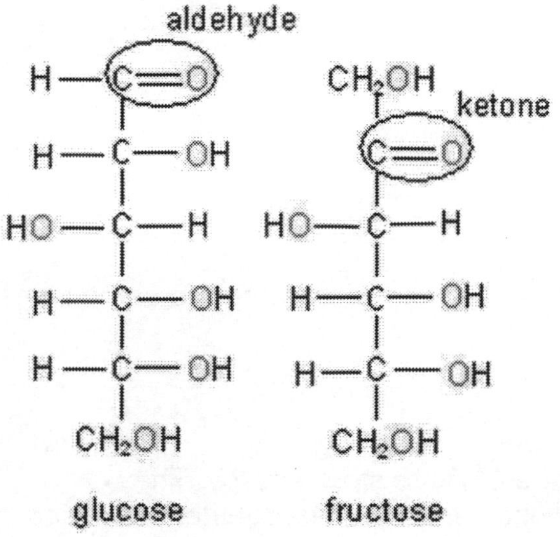

glucose   fructose

Aldehydes and ketones are organic compounds which incorporate a carbonyl functional group, C=O. The carbon atom of this group has two remaining bonds that may be occupied by hydrogen or alkyl or aryl substituents. If at least one of these substituents is hydrogen, the compound is an aldehyde. If neither is hydrogen, the compound is a ketone.

Glucose and fructose are the same in number of C, H, and O. One can see that aldehyde is the same as ketone in composition, but the location in glucose is different from the fructose.

Note that glucose and other kinds of sugars may be linear molecules or as in aqueous solution they become a ring form (hexagonal or pentagonal form).

Different shapes of molecules that are composed of the same number and kinds of atoms are called isomers. Glucose and fructose (shown below) are both $C_6H_{12}O6$ but the atoms are arranged in different configuration in each molecule, Figure (10).

There are two isomers of the ring form of glucose. They differ in the location of the OH and H groups.

Figure (10): Isomers of glucose

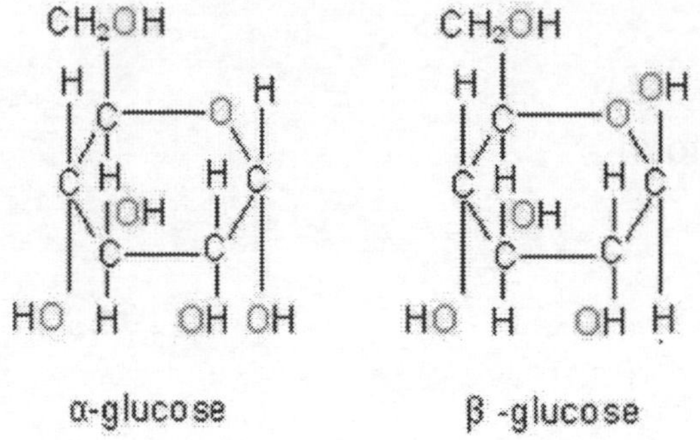

α-glucose   β-glucose

2-2 Polysaccharides

Polysaccharides are a chain of Monosaccharide molecules. Here are some types of polysaccharides:

2-2-1 Starch and Glycogen

The function of starch and glycogen, (Figure 11) is to store energy. They are composed of glucose monosaccharide (monomers) bonded together producing long chains.

Glycogen is stored in the liver and muscles. During fasting or between meals, the liver releases glycogen (by an enzyme called glucagon) in order to balance the sugar (glucose). Extra glucose is stored back in the liver as glycogen.

Plants produce starch to store carbohydrates and converted back to energy through the photosynthetic process.

Figure (11): Glycogen or starch

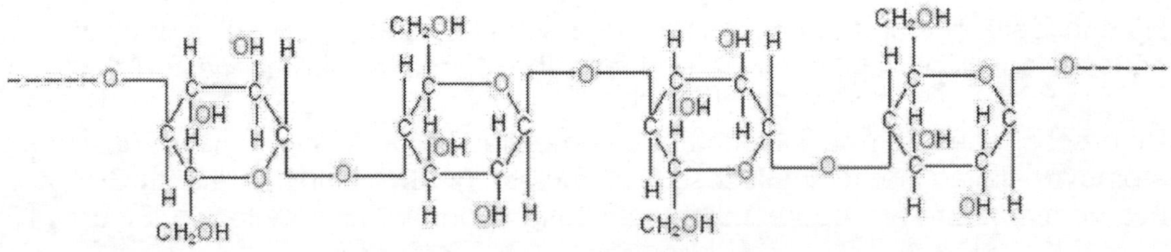

2-2-2 Cellulose and Chitin polysaccharides

Cellulose supports and protects the cell walls of plants. The cell walls of fungi and the exoskeleton of arthropods are composed of chitin, Figure (12). Cellulose is shown in Figure (13).

Figure (12): Polysaccharides of Chitin

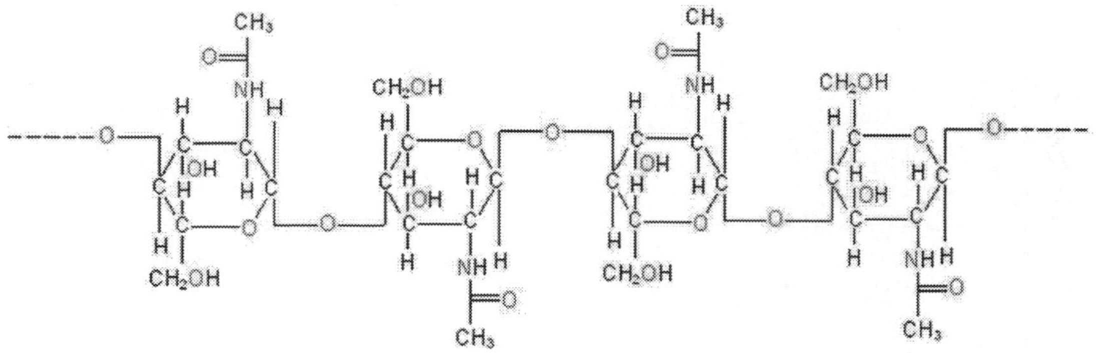

Figure (13): Polysaccharides of Cellulose

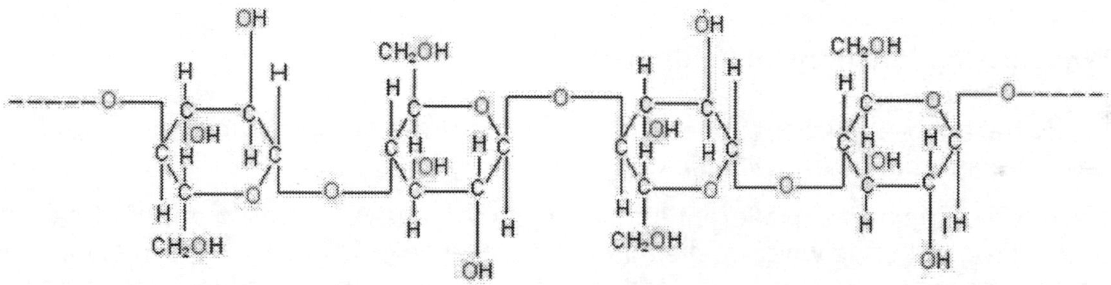

With energy from light (photosynthesis), plants can build sugars from carbon dioxide and water. Glycerin (also called glycerol) is not a sugar but is basically one half of a glucose sugar.

Humans and most animals do not have the necessary enzymes needed to digest the cellulose or chitin. Animals that can digest cellulose often have microorganisms in their gut that digest it for them. Fiber is cellulose, an important component of the human diet.

With energy from light, plants can build sugars from carbon dioxide and water. Glycerin (also called glycerol) is not a sugar but is basically one half of a glucose sugar, Figure (14).

Figure (14): Glucose and Glycerin.

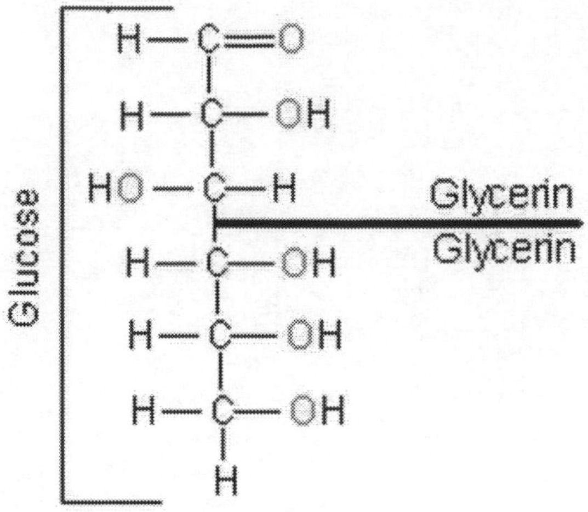

What are the functions of lipids (fat)?

Lipids have three functions; energy storage, forming the membrane around our cells, and hormones and vitamins.

Each type of lipid has a different structure and they have a large number of carbon and hydrogen bonds which makes them non lone polar group of molecule. Oxygen which is stronger than carbon in separating and pulling the hydrogen from carbon makes lipids very energy-rich.

As we know that water has two lone poles, the atoms are too strong to be separated from each other, and accordingly they do not attached to the carbon atoms in the lipid. Thus lipids are insoluble in the water, and they are stored in our body which has a large amount of water.

Most lipids are composed of some sort of fatty acid arrangement. The fatty acids are composed of methylene (or Methyl) groups and carboxyl (acid) as shown in Figure (15) below.

Figure (15): Unsaturated and saturated fatty acid

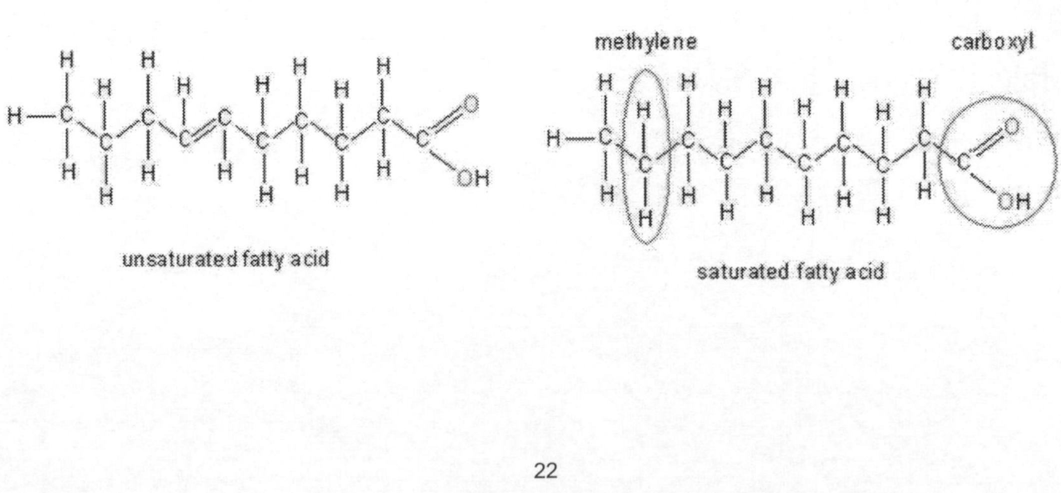

unsaturated fatty acid

saturated fatty acid

Fatty Acids: Acid means (–OH). Unsaturated fatty acid is a chain of Methylene with at least two carbon atoms lost their hydrogen atoms, as shown in Figure (15).

The fatty acid chains are usually between 10 and 20 Carbon atoms long. The fatty "tail" is non-polar (Hydrophobic; hates water) while the Carboxyl "head" is a little polar (Hydrophilic, loves water).

A fat is a solid at room temperature, while oil is a liquid under the same conditions. http://bioweb.wku.edu/courses/biol115/Wyatt/Biochem/Lipid/lipid1.htm - #The fatty acids in oils are mostly unsaturated because they have less hydrogen which is the lightest in the periodic table, while those in fats are mostly saturated, and therefore float such as butter and margarine.

The double bond also gives unsaturated fatty acids a strong bond (denser) in the methylene chain. And stick to each other. These interactions make them less fluid and more solid.

What are the types of lipids?

They are:

Fatty Acids
- Saturated
- Unsaturated
Glycerides
- Neutral
- Phosphoglycerides
Complex Lipids
- Lipoproteins
- Glycolipids
Nonglycerides
- Sphingolipids
- Steroids
- Waxes

Figure (16): Monoglycerides, diglycerides, or triglycerides

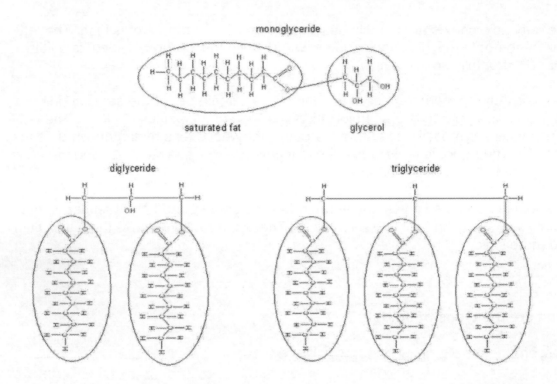

Extra triglycerides are converted into glugacon and then to fatty acid in the liver and to glycerol for the brain, Figure (17).

Figure (17): Triglycerides converted to Fatty acid and glycerol

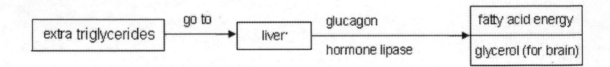

What is the protein?

Amino acids contain both a carboxyl group (COOH) and an amino group (NH$_2$). The general formula for an amino acid is given in Figure (20) below. Amino acids are either polar or non polar charges. The acidic positive COOH (COO$^-$) and the negative NH3 (NH3$^+$) cancel each other in some amino acids. Some amino acids molecules have larger charge in one side than the other and polar bonds.

Figure (18) Polar and non-polar amino acid

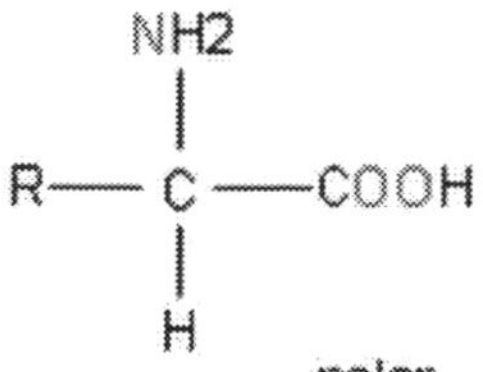

```
            NH2
             |
     R ——— C ——— COOH
             |
             H        polar
```

R is the functional group of the amino acid, and can be replaced by H

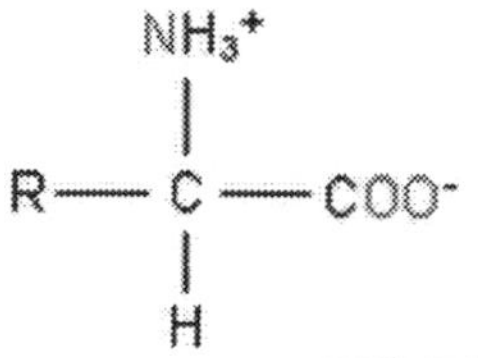

```
            NH3+
             |
     R ——— C ——— COO-
             |
             H        non-polar
```

Proteins have 20 amino acids and have different combinations of chains of amino acids.

What are types of fatty acids?

They are:

Type of fatty acid	
Fatty Acids	Glycerides
Saturated Fatty Acids	Triglycerides
Unsaturated Fatty Acids	Phosphoglycerides
Soap (salt of fatty acid)	
Prostaglandins	
Non glyceride Lipids	
Waxes	Steroids
Sphingolipids	Lipoproteins

What are esters?

The product of a condensation reaction in which a molecule of an acid unites with a molecule of alcohol with elimination of a molecule of water as shown in the following equation:

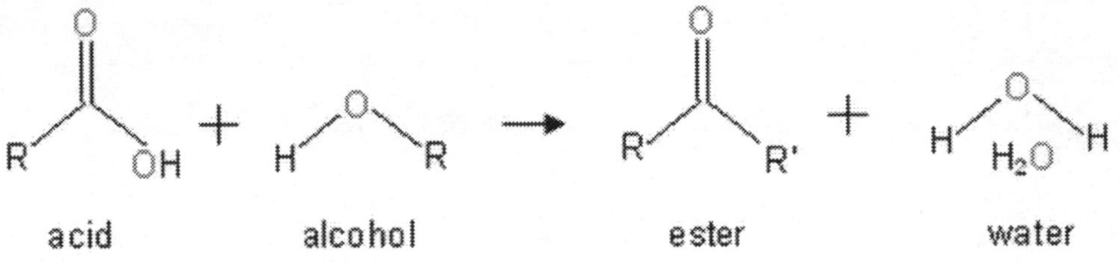

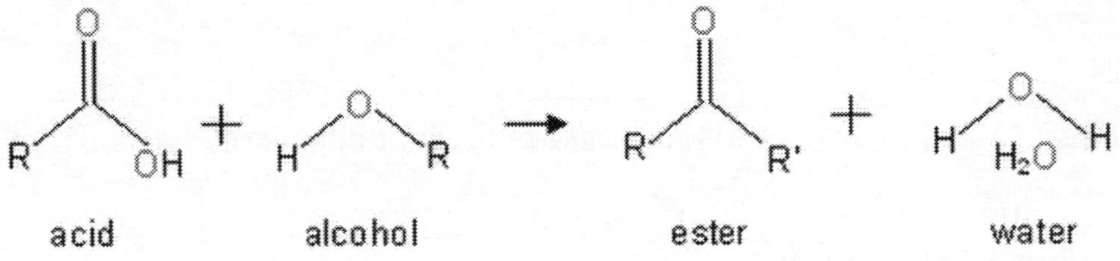

Some other types of esters are shown in Figure (19).

Figure (19): esters of carboxylic acid, nitric acid, phosphoric acid, and sulfuric acid

| carboxylic acid ester | nitric acid ester | phosphoric acid ester | sulfuric acid ester |

Esters can be represented in many forms. Esters can also be in the form of methyl propanoate (3 carbons), ethyl ethanoate, ethanoyl chloride, propanamide, and hydroxypropanentrite acids, as shown in Figure (20).

Figure (20): Different ester acids

Structure	Description	Name
$CH_3-CH_2-C(=O)-O-H$	Propanoate combines with methyl (CH3) which is alkyl group.	Methyl propanoate acid
$CH_3-CH_2-C(=O)-O-CH_3$	This stronger methyl propanoate replaces H with additional methyl group.	Methyl propanoate acid
$CH_3-C(=O)-O-CH_2-CH_3$	Ethyl group is C2H5 or CH2-CH3. Methyl is connected to ethyl group through COOH, and the acid has the characteristic of ethylene	Ethyl ethanoate acid
$-C(=O)-Cl$ $CH_3-C(=O)-Cl$	The one on the left is acyl chloride, and the one on the right is ethnoyl chloride.	Ethanoyl chloride acid
$CH_3-CH_2-C(=O)-NH_2$	Propane and ester amino group	Propanamide acid
$CH_3-CH(OH)-C\equiv N$	Ol, ide, and ate means 1, 2, 3 or more oxygens. This is called hydroxypropanentrite and not hydroxyproparentole. See IUPAC nomenclature later in this chapter.	Hydroxypropanentrite acid

$CH_3-CH2-NH_2$	Amine is NH3, and connected to ethyl group of C2H5.	Ethylamine acid
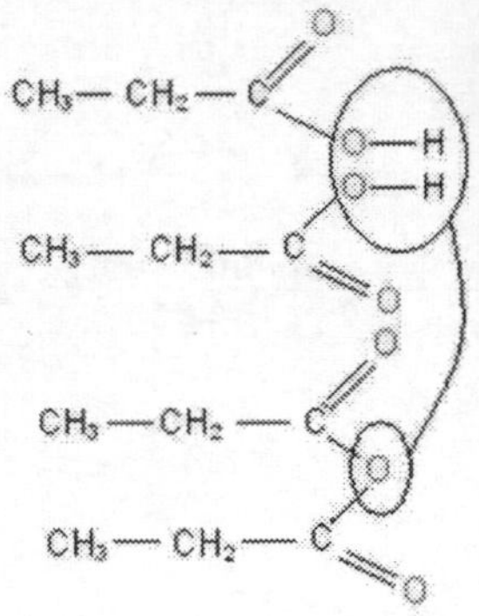 CH_3-CH_2-C (with $O-H$, $O-H$)	Anhydride means one molecule of water is taken out. The result is propanoic acid without water.	Anhydride propanoic acid
CH_3-CH_2-C ... CH_3-CH_2-C		
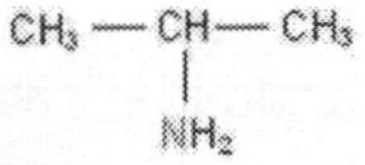 $CH_3-CH-CH_3$ with NH_2	This molecule has a propane molecule of which the second carbon is connected to amino group.	Amino propane acid
CH_3, NH, CH_3		Dimethylamine acid
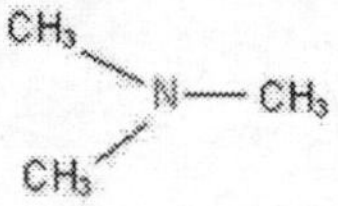 CH_3, $N-CH_3$, CH_3	Can you name this three methyle groups?	Trimethylamine acid
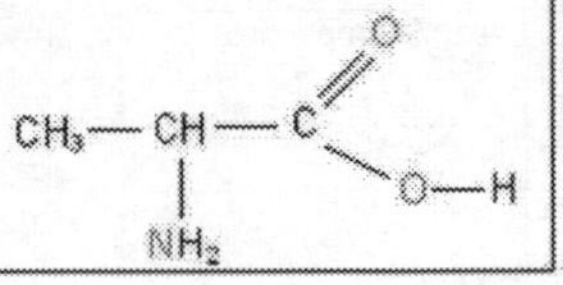 CH_3-CH-C with $O-H$ and NH_2	This is propane molecule with amine group.	Aminopropanoic acid (alanine)

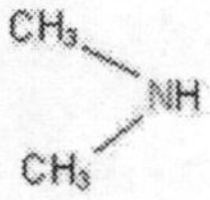

Structure	Description	Name
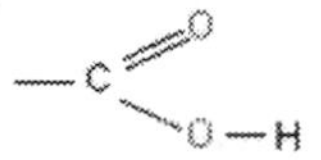 (—C(=O)—O—H)	Carboxylic acid contains -COOH.	Carboxylic acid
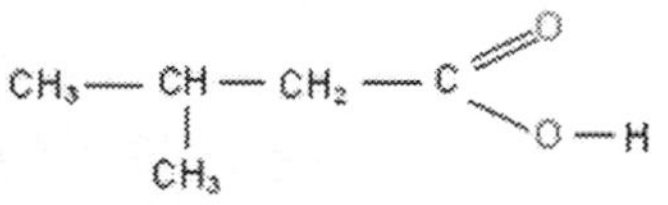 CH_3—CH—CH_2—C(=O)—O—H, with CH_3 branch	Methylbutanic acid has methyl group (-CH_3) which is the third carbon from the carboxylic carbon and called 3-methylbutanic acid.	Methylbutanic acid
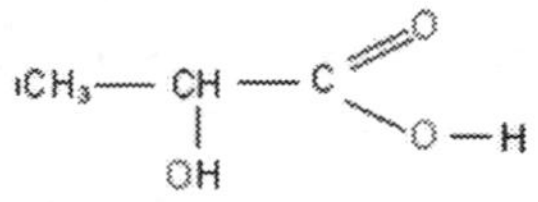 CH_3—CH—C(=O)—O—H, with OH branch	2-hydroxypropanoic acid in which the hydroxyl group comes on the second carbon from the carboxylic carbon. This is called lactic acid.	Hydroxypropanoic acid or lactic acid
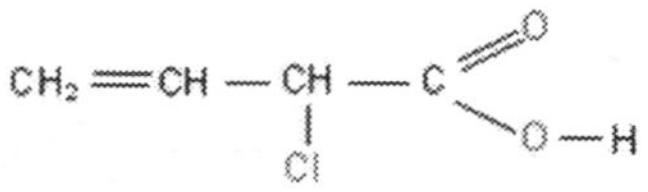 CH_2=CH—CH—C(=O)—O—H, with Cl branch	Chlorine atom is connected to the second carbon.	Clorobutanic acid
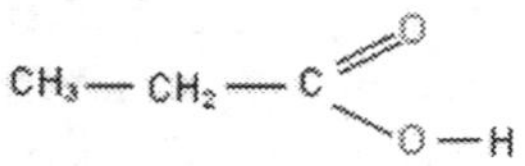 CH_3—CH_2—C(=O)—O—H	Propanoic acid is derived propane (3 carbons).	Propanoic acid
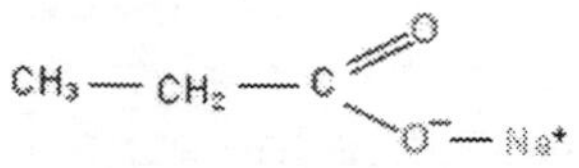 CH_3—CH_2—C(=O)—O^- — Na^+	Sodium (2, 8, 1) is weaker than oxygen (2, 6). Thus oxygen takes one electron from sodium atom. This is not acid because it does not have OH group.	Sodium propanoate salt

What are steroids?

A steroid is a chain of three hexa-rings and one penta-ring of carbon. Steroids vary by the functional group attached to these rings. Functional groups of O and OH and their location can create hundreds of steroids. Figure (21) shows different steroids molecules of different function in human. Steroids are also made in plants and fungi. Estrogen and progesterone are made in the ovary and placenta before and after the 15 days of the menstrual period respectively Testosterone is made in the testes, and some of it is also made in female to control the estrogen. Cholesterol in the mitochondrion is converted to the required steroid. Estrogen is the main component of contraceptive and hormone replacement therapy (HRT). Estrogen has OH at the end, and androgen has O at the end of the steroid molecule. Both men and women have both estrogen and androgen. In male, androgen is synthesized rapidly and same thing for the estrogen in female. After menopause, the estrogen is gradually reduced and the androgen is emerged to produce hair in older women. Some estrogens are also produce in the liver, adrenal gland and the breasts to control the mode of postmenopausal women. HRT is given to postmenopausal women to prevent osteoporosis. Estrogen controls the HDL and LDL cholesterol. Androgen controls the sex desire in both male and female. Hormone replacement therapy (HRT) should not be given after the start of pregnancy, because the level of progesterone is higher than the estrogen, otherwise the risk of cancer could be positive. Once the cancer established in the breast, the cancer loves the estrogen hormone. These cancers can be treated by suppressing the estrogen.

Figure (21): Androgen and estrogen steroids

androgen/progesterone

androgen/progesterone

steroid

What is the difference between Oxidation and reduction?

Oxidation and Reduction

Oxidation is defined as the addition of oxygen, and the reduction is the removal of oxygen. Oxidation and reduction can also be defined as the loss and gain of electrons respectively. The loss of electrons from an atom produces a positive oxidation state, while the gain of electrons results in negative oxidation state.

From the periodic table, one can simply see that oxidation is associated with metals, and reduction is associated with nonmetals. All metal atoms are characterized by their tendency to lose one or more electrons, forming a positively charged ion (oxidation), called a cation. During this oxidation reaction, the oxidation state of the metal always increases from zero to a positive number, such as "+1, +2, +3...." depending on the number of electrons lost from the metal or gained by the nonmetal. When the nonmetal gains a negative charge (electrons), it is called anion. See the table below.

Group Number	Number of electron lost or gained (L $ G)	Charge cation or anion (C & A)
I Hydrogen	1 L	+1 C
II Beryllium	2 L	+2 C
III Boron	3 L	+3 C
IV Carbon	4 L or 4 G	+4 C or - 4 A
V Nitrogen	3 G	-3 A
VI Oxygen	2 G	-2 A
VII Fluorine	1 G	-1 A
VIII Neon	0 Stable	0 Neutral

Number of electrons in the outermost orbit is often called valence electrons.

Electronegativity in metals is higher than in nonmetals. It is therefore that nonmetal atoms lose electrons to metals atoms. The table below shows negativity in some metal and nonmetal atoms.

Metallic elements							Nonmetallic elements			
Li	Be						C	N	O	F
1	1.5						2.5	3	2.5	4
Na	Mg	Al						P	S	Cl
1	1.2	1.5						2.1	2.5	3
K	Ca	Sc	Fe	Ni	Cu	Zn			Se	Br
0.9	1	1.3	1.8	1.9	1.9	1.7			2.5	2.8

By convention oxidation and reduction reactions are written in the following form:

Atom   Number of electrons   Atom charge

X   + $n (e^-)$ X^{-n}

X   - $n (e-)$ X^{+n}

The carbon for example has two states:

C   + $4e^-$ C^{-4}
C   + $4 (-e)$ C^{+4}

As the nonmetal atom gains the electrons lost by the metal, it reduces its state of negativity from zero to a negative value (-1,-2.-3) depending on the number of electrons gained by the nonmetal, Figure (22).

Figure (22): Oxidation and reduction in metal and nonmetal

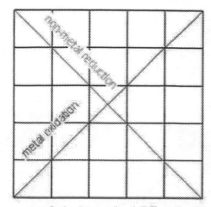

no. of electrons (gain)/(loss)

Note, the GROUP VIII nonmetals have no tendency to gain additional electrons, hence they are non-reactive in terms of oxidation-reduction. This is one the reasons why this family of elements was originally called the inert gases.

Oxidation-reduction reactions mean that the process of oxidation cannot occur without a corresponding reduction reaction. Oxidation must always be "coupled" with reduction, and the electrons that are "lost" by one substance must always be "gained" by another, Figure (23).

Figure (23): Oxidation of some molecules

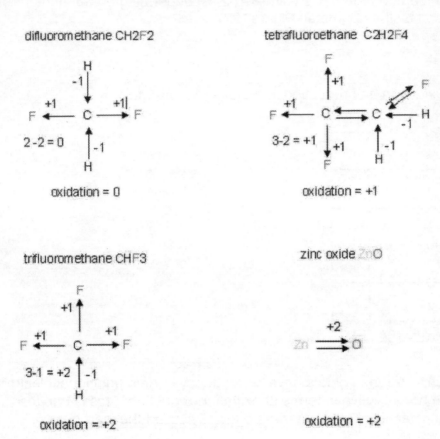

difluoromethane CH2F2
H
-1
+1
+1
F
C
F
2-2 = 0
-1
H
oxidation = 0
tetrafluoroethane C2H2F4
F
+1
F
+1
F
C
C
H
-1
3-2 = +1
+1
-1
F
H
oxidation = +1
trifluoromethane CHF3
F
+1
+1
+1
F
C
F
3-1 = +2
-1
H
oxidation = +2
zinc oxide ZnO
+2
Zn
O
oxidation = +2

methan CH₄

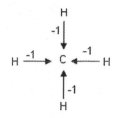

H
-1
H —-1→ C ←-1— H
-1
H

oxidation = -4

ethane C₂H₆

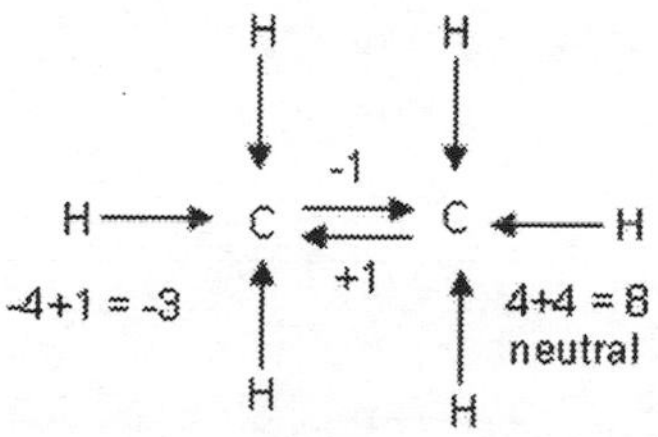

H   H
-1
H —→ C ⇄ C ←— H
+1
-4+1 = -3   4+4 = 8
neutral
H   H

oxidation = -3

methyl fluoride CH₃F

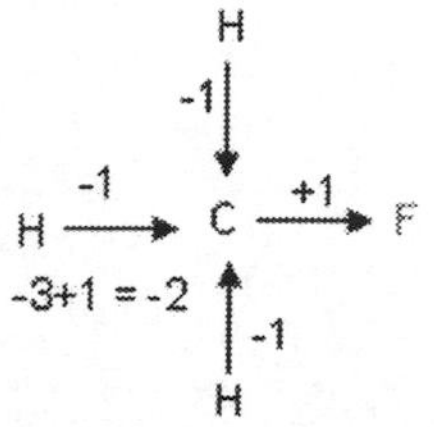

H
-1
H —-1→ C —+1→ F
-3+1 =-2
-1
H

oxidation = -2

acetylene C₂H₂

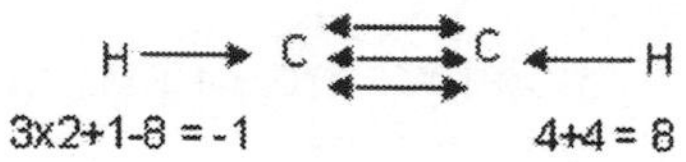

H —→ C ⇄⇄ C ←— H
3x2+1-8 = -1   4+4 = 8

oxidation = -1

potassium dioxide KO₂

-1/2
K ⇒ O2

oxidation = -1/2

oxygendifluoride OF₂

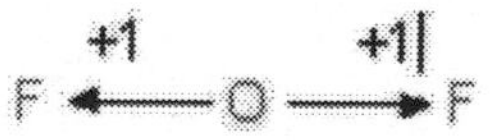

+1   +1
F ←— O —→ F

oxidation = +2

rubidium trioxide RbO₃

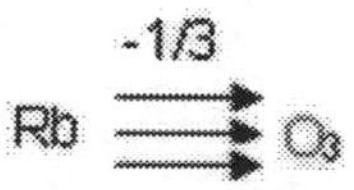

-1/3
Rb ⇛ O₃

oxidation = -1/3

oxygen O₂

0
O ⇄ O

oxidation = 0

Oxidation and reduction in terms of Oxygen

- Oxidation is gain of oxygen
- Reduction is loss of oxygen

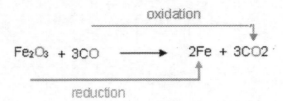

The iron oxide is called oxidizing agent, and the carbon monoxide is called reducing agent. In other word.

Reduction and oxidation are known as a redox reaction.

Oxidation and reduction in terms of hydrogen transfer

o Oxidation is loss of hydrogen.
o Reduction is gain of hydrogen.

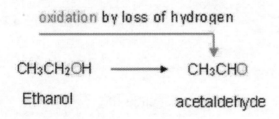

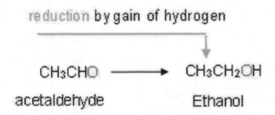

o Oxidizing agents give oxygen to another substance.
o Reducing agents remove oxygen from another substance

$$\overset{0}{Cu} + 2Ag\overset{+1}{N}\overset{+5}{O}\overset{-2}{3} \longrightarrow \overset{+2}{Cu}(\overset{+5}{N}\overset{-2}{O3})_2 + 2\overset{0}{Ag}$$

$$2 \times (+1 + 5 - 2) \qquad 2 + (5 - 2)_2$$
$$2 \times 4 = 8 \qquad 2 + 6 = 8$$

Note: O3 is considered as one oxygen, see the diagram below:

2AgNO3

Cu (NO3)2

Copper in the above equation (left hand side) as has zero charge, as it stands alone. The Oxygen in AgNO3 is the most negative atom, and therefore it tends to take all electrons from Ag and N. Therefore, the Oxygen will have (-2) charge, N will have +5, and Ag will have (+1).

The left side of the equation will have Cu (+2), N (+5), and O (-2). If we equate both sides of the equation, then both sides are equal to zero charge.

Examples

Sulfur trioxide

$$2S + 3O2 \longrightarrow 2SO3$$
$$0 \qquad 0 \qquad +6\ -2$$

Note that both oxygen and sulfur ends with 6 electrons in the outermost orbit, but the oxygen is more negative, because the second orbit in oxygen is 6 and the third orbit in sulfur is also six.

Water

$$2H_2 + O_2 \longrightarrow 2H_2O$$
$$0 \quad\quad 0 \quad\quad\quad\quad +1 \ -2$$

Combination reaction can be reversed, i.e. a compound can be decomposed into the components from which it was composed. This type of reaction is called decomposition reaction.

Potassium chloride

$$2KClO_3 \longrightarrow 2KCl + 3O_2$$
$$+1 \ -1 +2 \quad\quad\quad +1 \ -1 \quad\quad 0$$

Redecompositioning can not happen in some reactions.

Examples

Calcium Carbonate

$$CaCO_3 \longrightarrow CaO + CO_2$$
$$+2 +4 -2 \quad\quad\quad +2 -2 \quad +4 -2$$

Note that O3 has only -2 because the carbon atom gets its eight electrons from calcium (2 electrons), Oxygen (two electrons), and the existing four electrons in its outside orbit, as shown in the molecular formula of $CaCO_3$ below:

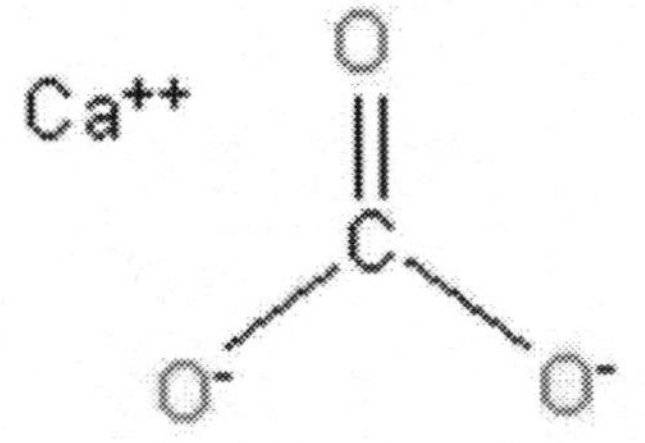

Iron and hydrochloric acid

$$2Fe + 6HCl \longrightarrow 2FeCl_3 + 3H_2$$

$$\begin{array}{ccccc} 0 & +1 & -1 & +2 & -1 & 0 \end{array}$$

Oxidation States of the elements

Chemical elements tend to have their outermost orbits in stable conditions,i.e. 2, 8, 18, and 32. Let's take some examples:

Beryllium (2, 2)

Beryllium has 2 electrons in its inner orbit and 2 in the outside orbit. It tends to give the outside two electrons to those elements which are more negative. It is difficult to gain electrons as the outside orbit is much far from the number 8. Thus, the oxidation state is +2.

Carbon (2, 4)

The outermost orbit is 4, and therefore it needs 4 electrons to complete the orbit to 8, or it gets rid off 4 electrons to keep 2 electrons in the outermost orbit. Carbon oxidation states can be written as -4, -3, -2, -1, +1, +2, +3, +4 accordingly.

Nitrogen (2, 5)

Nitrogen oxidation state can be written as -3, -2, -1 N +1, +2, +3, +4, +5. The table below shows oxidation states for more elements.

Oxygen (2,6)	-2, -1 O +1, +2
Na (2,8,1)	-1 Na +1
Al (2,8,3)	Al +1,+3
Si (2,8,4)	-4,-3,-2,-1 Si +1, +2,+3,+4
Cl (2,8,7)	-1 Cl +1,+2, +3, +4, +5, +6, +7
Ti (2,8,10,2)	-1 Ti +2, +3, +4 Note that 10 electrons is not a stable number, so it can accept any number of electrons less than 5 which is (10 + 2 - 4) = 8
Cr (2,8,13,1)	-2, -1 Cr +1, +2, +3, +4, +5, +6 Note that -2 makes13+1+2 - 8 = 8, and +6 makes 13 + 1 + 6 - 8 = 8
Mn (2,8,13,2)	-3, -2, -1 Mn +1, +2, +3, + 4, +5, +6, +7
Fe (2,8,14,2)	-2, -1 Fe +1, +2, +3, +4, +5, +6
Zn (2,8,18,2)	Zn +2, Zn Can not accept more electrons to keep It orbit stable (18 =2+10+8), and can not give more than 2 electrons.
Mo (2,8,18,13,1)	-2, -1 Mo 1, +2, +3, +4, +5, +6. This metal has very weak outermost orbitals and can easily accept or give electrons to other atoms.
I (2,8,18,18,7)	-4, -3, -2, -1 I +1, +3, +5, +7
At (2,8,18,32,18,7)	-4, -3, -2, -1 At +1, +3, +5, +7
U (2,18,32,21,9,2)	+3, +4, +5, +6

Let's see the oxidation and the reduction of some atoms in the table below.

-4	-3	-2	-1	Element	1	2	3	4	5
			-1	H	1				
			-1	Li	1				
			-1	Na	1				
				K	1				
-4	-3	-2	-1	C	1	2	3	4	
-4	-3	-2	-1	Si	1	2	3	4	
	-3	-2	-1	Na	1	2	3	4	5
	-3	-2	-1	P	1	2	3	4	5
		-2	-1	O	1	2			
		-2	-1	Si	1	2			

One can see from the above table that the nitrogen and the phosphorous have the tendency for oxidation more than other elements. Nitrogen and Phosphorous are vital elements in the formation of amino acids and the DNA.

As a general rule, atoms with less electronegativity tend to be oxidized more than atoms with higher negativity. Figure (24) has the first three rows of the last five elements of the periodic table, namely;

B C N O F
Al Si P S Cl
Ga Ge As Se Br

Figure (24): Negativity of previous elements

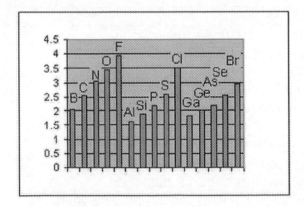

Figure (24): Continued

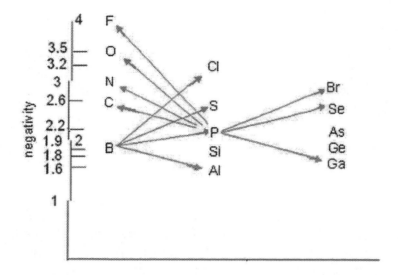

One can see that B (Boron) is less negativity than P (Phosphorous), S (Sulfur), and Cl (Chlorine), and therefore will be oxidized if it is bonded with any of them. However, it will be reduced if it is bonded with aluminum, as the later is lower in negativity. Boron is neither in negativity nor in positivity with Si (Silicon). Similarly, P (phosphorus) is oxidized with respect to C (carbon), N (Nitrogen), O (Oxygen), and F (Fluorine). Ga (Gallium) is oxidized with respect to Phosphorus. The Phosphorus has the tendency for oxidation more than any other element in Figure (24).

What are isotopes?

An isotope is an element whose nucleus contains a specific number of neutrons, in addition to the number of protons that uniquely defines the element. It means that an atom of the same element can have different number of neutrons and a fixed number of protons. An atom with different number of neutrons are called isotope. Hydrogen has no neutrons at all. However there are some hydrogen isotopes; deuterium (hydrogen with one neutron), and tritium with two neutrons, Figure (25).

Figure (25): Isotopes of hydrogen

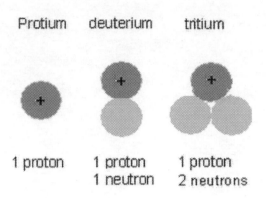

Protium   deuterium   tritium

1 proton   1 proton   1 proton
              1 neutron   2 neutrons

Isotopes are written in the form $^{A}B_{C}$, where B is the symbol of the element, C is the atomic number, and A is the number of neutrons and protons combined, called the mass number. For example, ordinary Hydrogen has the formula of $^{1}H_{1}$, deuterium $^{2}H_{1}$, and tritium is $^{3}H_{1}$. Every chemical element has more than one isotope, and one isotope is more abundant in nature than any of the others. Multiple isotopes of one element can be found mixed.

Certain isotopes of elements are unstable, radiating energy or radioactive waves, and called radioisotopes. Carbon - 14 ($^{14}C_{6}$) is a radioisotope. Certain isotopes are more radioactive than others. There are many carbon isotopes ranging from 8C to 22C. All of them have halftimes ranging from several seconds to microseconds, except the 14C which has 5730 years of halftime. Radioisotopes have an unstable nucleus that emits rays of types alpha, beta, or gamma (will be explained) until the element reaches stability point.

The stability point in this case is a nonradioactive isotope of another element. For example, radium - 226 ($^{226}Ra_{88}$) will decays finally to lead - 207 ($^{207}Pb_{82}$). Precise measurement shows that some elements still contain traces of radioisotopes even with light elements such as hydrogen. Isotopes of phosphorous are used in medical therapy to kill certain cells such as cancerous cells such as leukemia (increase in white blood cells) and polycythemia (increase in red cells).

In medical diagnosis and research, isotopes of iodine are used in the diagnosis of thyroid function and in the treatment of hyperthyroidism, and in cathodic - ionic blood circulation action and reaction of substances for research. In industry, radioisotope is used in measuring the thickness of metal or plastic sheets. Testing of corrosion, oxidation, reduction, thickness of paint and other industrial process can be tested by radioisotopes.

Radioisotope releases energy (decay) by spitting out energy in the form of particles or electromagnetic waves.

What Types of Radiation Are There?

The radiation one typically encounters is one of four types: alpha radiation, beta radiation, gamma radiation, and x radiation.

Radioisotopes are continually undergoing decay.

Alpha decay is common with elements of atomic number greater than 83; from rubidium onwards. Alpha decay can be written in the following form:

$$^AB_C \longrightarrow {}^AB_C - {}^4He_2 \longrightarrow {}^{A-4}D_{C-2}$$

Where B is the parent isotope and D is the daughter isotope and different element than element B. Helium (He) is called the alpha particle.

The new element D is now less in atomic number by 2, and consequently going back two places in the periodic table. Can we convert thallium (81) to gold (79)?

Helium has two electrons in the first orbit, and they are strongly attached to the nucleus which has two positive protons. The two electrons are very difficult to leave the helium atom, yet they love to capture electrons. Therefore, alpha is not very penetrating as it captures electrons immediately from the substance it hits. However, it is very damaging because the energy released, due to the capture of more electrons, can knock atoms off the substance.

Alpha rays are used in many applications. Since alpha decay travels short distance (few centimeters or inches), it can be used to ionize small air (mainly CO_2) and allow a small current to flow such as smoke detectors. CO_2 will be ionized to CO_2^- and a current of several milliamps go through the gap of the smoke detector, sounding the alarm.

Alpha decay is used in thermoelectric generators used for space probes and artificial heart pacemakers, because alpha decay is much more easily shielded against than other forms of radiations. Plutonium -238 (the product of plutonium 244 : $^{244}Pu_{94} - {}^4He_2 = {}^{238}Pu_{92}$) requires only one inch (2.54 centimeters) of lead to protect against other radiations.

Beta negative decay is the process of converting the neutron into a proton, and it follows the equation:

$$^AB_C \longrightarrow {}^AB_C - {}^0e_{-1}{}^{-1} \longrightarrow {}^AB_{C+1}$$

The tritium (H-3) can be converted into helium, Figure (26), by converting one neutron into a proton. So $^3H_1$ becomes $^3H_2$, and an unstable isotope is converted into a stable isotope of helium, or semi stable helium. For complete stability, the number of protons must equal to the number of neutrons.

Figure (26): Beta negative decay turns tritium into helium

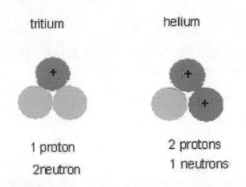

Beta negative decay is fast and furious at the beginning and slow down over time. It is more penetrating than alpha because the particles are smaller.

In general, there are three forms of beta decay:

a) Electron emission in which a neutron converts into a proton with the emission of an electron which is called a beta-minus particle. It takes the form:

$$A^0 \longrightarrow P^+ + e^-$$

Note that the total charge is the same on both sides of the formula. The electron e⁻ can not be existed inside the nucleus and therefore it is repelled and released to outside the nucleus.

When the electron is repelled, it could cause damage to the cell of a human body. So if beta decay is not properly shielded, the electron could oxidize the human cell and cause mutation to it, and possibly a cancerous result. An example of this negative decay occurs in the iodine -131 which decays into xenon - 131 and one electron:

$$^{131}I_{53} \longrightarrow {}^{131}Xe_{54} + {}^0e_{-1}$$

Note that the alpha decay decreases the atomic number, whereas the beta decay increases the atomic number. In other words, Alpha is a reducing agent, and the beta is oxidizing agent. Above equation shows that the number of protons have been increased, and if the number of protons have increased above a certain limit, the atom will not be stable. In this case, the atom attempts to be stable again by converting some protons back to neutrons with the emission of a positively - charged electrons.

b) An electron with a positive charge is called a positron:

$$P^+ \longrightarrow n^0 + e^+$$

An example of this type of decay potassium - 39 which decays into Argon -39:

$$^{39}K_{19} \longrightarrow {}^{39}Ar_{18} + {}^{0}e_{+1}$$

c) Electron capture or beta decay X-ray, in which an inner orbiting electron is attracted into unstable nucleus and combines with a proton to form a neutron:

$$e^- + p^+ \longrightarrow n^0$$

In this process, there is no radiation emitted, but the cloud surrounding the nucleus is changed, and filled by an electron from an outer shell. The filling of the vacancy is associated with the emission of X-ray. That is called beta decay of X -ray type.

Gamma decay involves the emission of energy from an unstable nucleus in the form of electromagnetic or photon radiation. It follows the following formula:

$$^{A}B_C \longrightarrow {}^{A}B_C + {}^{0}\gamma_0$$

In gamma decay, a nucleus changes from a higher energy state to a lower energy state through the emission of electromagnetic radiation or photons. The number of protons (and neutrons) in the nucleus does not change in this process, so the original and the output atoms are the same chemical element, Figure (27).

Figure (27): Gamma change in energy

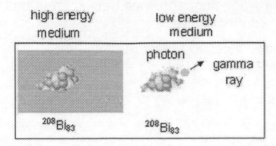

In alpha and beta decay, the atomic number of the nucleus changes, but in gamma decay the atomic number does not change. In gamma decay, the change is only in the energy state of the nucleus that will be changed to a lower state by emitting photons. The photons produced in this decay are consequently known as gamma rays and have a wavelength with an order of magnitude of about 1,000 $\times 10^{-15}$ (10^{-15} = 1 femtometer),= 10^{-12} meters. As a result, the nucleus will decay to the ground state (ground state is the condition of an atom, ion, or molecule, when all of its electrons are in their lowest possible energy level i.e., not excited. Ground state also means that all electrons fill the lowest energy orbits.) by emitting one or more gamma-ray photons, Figure (28).

Figure (28): Comparison between the three isotopes in penetration

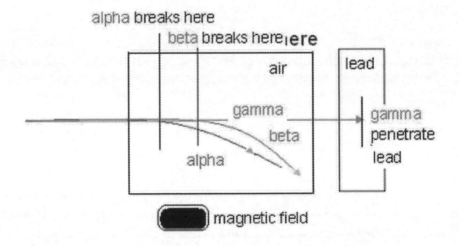

Alpha and beta decays can change the nucleus structure of an atom, i.e., the atom will be in an excited state after the decay is completed. After alpha and beta decay, gamma decay will take responsibility by releasing more energy until the atom reaches the ground state, Figure (29).

Figure (29): Staging of radioactive

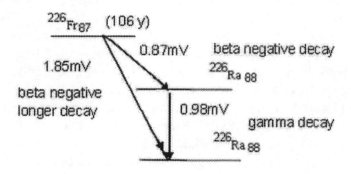

We shall show full staging of energy and atomic number for the three decays; alpha, beta negative, beta positive, gamma in Figure (30).

Figure (30): Staging of alpha, beta negative, beta positive, and gamma decay

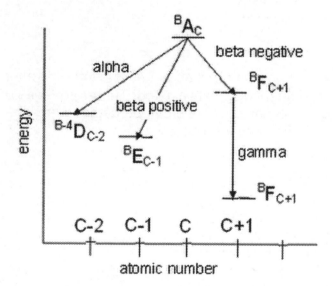

What does half-time mean?

Radioactive atoms (unstable isotopes) are steadily disappearing, being replaced by stable atoms. The decay from parent to daughter occurs at a certain rate. The rate of decay can be expressed as the time it takes for half of the weight of the parent to convert to its daughter. For example, the half time of beryllium 11 is 13.81 second, and let's start with 20 grams of ^{11}Be. After 13.81 seconds, the weight if beryllium becomes 10 grams, and the rest is converted to boron - 11. Another 1.81 seconds, we will have 4 grams of beryllium -11, and 2 grams after 13.81 seconds more, and so on, until the beryllium -11 is vanished. We are talking about Be - 11 not Be - 9.012 which is stable. All radioactive elements disintegrate according to their specific half time, or half life. The half time of a radioisotope is the time required for half of the initial mass number of an atom to disintegrate. The following formula represents the decay time and the atomic mass of an atom:

$$Ln \frac{M}{Mo} = -kt$$

Where M is the remaining weight, Mo is the initial weight, k is the rate of decay, and t is the time of decay. Solve above equation, considering M/Mo equals 0.5, then ln o.5 = - 0.693. Therefore, the half time equation can be written as;

$$t_{\frac{1}{2}} = \frac{0.693}{k}$$

Using above equations, one can predict how much of an element is left over after a certain time, and how much of the element originally existed. Therefore, ages of archeological history of a substance, material, rocks, etc can be determined. The table below shows some element with their half time decay.

Uranium	^{226}U$_{92}$	4.5×10^{9}	y
Radium	^{226}Rn$_{88}$	1602	y
Radon	^{222}Rn$_{86}$	3.82	y
Astaline	^{210}At$_{85}$	8.1	h
Francium	^{223}Fr$_{87}$	22	min
Ununbium	^{285}Uub$_{112}$	34	sec
Ununhexium	^{293}Uuh$_{116}$	5.3×10^{-2}	sec

Example: Assume there is 875 grams of radon for every 125 grams of radium as:

1000 grams of radium (initially) $\rightarrow$ 500 $\rightarrow$ 250 $\rightarrow$ 125 grams of radium

Therefore, 1000 -125 = 875 grams of radon occurred in three half times.

The proportion of mass equals to 875 divided by 125 which equals to 7.

Thus 7 times 1602 years of half times gives 11214 years have elapsed since the 1000 grams of radium converted to 875 grams of radon.

What Are Acids, Bases, and pH all about?

The word pH stands for "power of hydrogen". It means that if a soluble compound has a lot of hydrogen, it is acidic, and with little hydrogen, it is a base (alkali). How can we reduce the level of hydrogen in a compound? It is simple, just by oxidizing it (bond it with oxygen) to become OH^-. The measurement of acids and bases is by the pH scale. The scale goes from very close to 0 through 14 as follows:

- Strong acid has a very low pH. (0 - 4)
- Strong base has a very high pH (10 - 14)
- Weak acid has a range "between" 3 - 6 on pH scale. It is partially ionizes in an aqueous solution (water).
- Weak base between 8 -10, and partially ionizes in water.
- Neutral has a pH of 7. It is neither acidic nor basic.

Acids are compounds that break into hydrogen H^+ ions and another compound when placed in aqueous solution. Bases are compounds that break into OH^- and other compounds. In other words, acids are hydrogen (H^+) donors (proton donors), and bases are oxidized hydrogen (OH^-), or hydroxide, donors. So, if an ionic compound is put in water, it will break into two ions. If the compound is acidic, more H^+ ions than OH^- will be released, and vice versa with the basic (alkaline) compound.

Fluids of the body cannot be too acidic (high level of H^+) and not too basic (high level of OH^-), and the body buffers the level of both acidity and alkalinity by producing carbonic acid ($H2CO3$) or bicarbonate ($HCO3$). If there is an excess of acidity (H^+), then the body exhale lots of carbon dioxide:

$$H^+ + HCO_3^- \rightarrow H2CO_3 \rightarrow H_2O + CO_2$$

So, one should not drink or eat lots of acidic food, otherwise he will suffer from hypoxia that could lead to pulmonary hypertension and asthma.

If there is a shortage of H^+ then:

$$H_2CO_3 \rightarrow H^+ + HCO3^-.$$

The interpretation of pH value is:

pH 7 = 10^{-7} mol H$^+$ per liter or 10^{-7} mol OH$^-$ per liter, and

pH x = 10^{-x} mol H$^+$ per liter or 10^{-x} mol OH$^-$ per liter.

Some pH values are listed in the table below.

Name	pH level	Molecular formula
Gastric juice	1.2 - 3.0	
Hydrochloric acid	0	HCl
Stomach acid	1	
Lemon Juice	2	
Vinegar	3	CH$_3$COOH
Veginal fluid	3.5 - 4	
coffee	5.0	
Milk	6 - 6.8	
Soda	4	Na2CO3, Na2O
Rain water	5	
Distilled water	7	H2O
Baking soda	9	NaHCO3
Blood	7.35 - 7.45	
Bile (Bilirubin)	7.6 - 8.6	C33H 36 N4O6
Pancreatic juice	7.1 - 8.2	
Tums (antacid) (calcium carbonate)	10	CaCo3
Ammonia	11	NH3
Mineral lime	12	Ca(OH)2
Drano (sodium hydroxide)	14	NaOH

Acids and basis can be neutralized by mixing them together to give water and salt:

HCl (acid) + NaOH (base) = H$_2$O (water) + NaCl (salt)

The question is why the oxygen, in above equation, gets rid of Na and takes instead the atom H. The answer is that the oxygen in NaOH has one dot of one pole is attached to hydrogen and the other dot is attached to Na, and the hydrogen of negativity of 2.1 repels the Na atom which is of negativity 0.9. Thus there is an angle of about 180 degrees between them, where as in H$_2$O the angle is about 104 degrees. Therefore, the repulsion between two hydrogen atoms is less than the repulsion between hydrogen and sodium, and consequently the hydrogen atom prefers the other hydrogen atom to bind.

Same interpretation happens with the following equation:

$$HBr \text{ (acid)} + KOH \text{ (base)} = H_2O \text{ (water)} + KBr \text{ (salt)}$$

It is important to understand the nature of acidity and alkalinity for the treatment of cancer. Cancer cells become dormant at a pH slightly above 7.4 and die at pH 8.5.

In other words, increasing alkalinity such as eating vegetation diets and drinking fresh fruit and vegetable juices can treat cancer.

The pH scale goes from 7 to 14, and the blood, lymph and cerebral spinal fluid in the human are designed to be slightly alkaline at a pH of 7.4. Cancerous cells can live without oxygen and love hydrogen, i.e. acidic environment, whereas healthy tissues are alkaline. When oxygen enters an acidic solution it can combine with H^+ to neutralize the acidity.

In Holland, the vegetation diet promoted by Dr. Moerman has been recognized by the government as a legitimate treatment for cancer, and the result indicated that vegetation diet is more effective than standard cancer treatment.

Is there any other term for the definition of acidity?

Yes. Acidity can also be dfined in terms of Ka and pKa

Let's take this example:

$$HCl \text{ (aq)} + H_2O \rightarrow H_3O^+ \text{ (aq)} + Cl^- \text{ (aq)}$$
$$\text{Acid} \qquad \text{base} \quad \text{cation} \qquad \text{anion}$$

$H3O^+$ is also called conjugate acid
Cl^- is also called conjugate base

$$Ka = \frac{[H_3O^+][Cl^-]}{[HCl \times H_2O]}$$

Note that the unit of numerators and denominators is concentration

$pKa = -\log_{10} ka$

Ka is called acid dissociation constant or acidity constant or acid - ionization constant. It is a measure of the strength of an acid in solution. The larger the value of Ka is the stronger the acid. Ka values cover a wide range of 10^{+10} for the strongest acid such as sulfuric acid to 10^{-50} for the weakest acid such as methane. The negative sign is more appropriate to be assigned to acidity as acidity is below pH 7, and the positive sign is for alkalinity. So, pK is a more convenient scale than

Ka. As an example, Ka of !0 +10 is vey acid solution, but if we say pKa is -10, it is more acceptable: pKa = -log $_{10}$ 10 = -10.

Similarly, pKa = -log $_{10}$ -50 = 50 is more appropriate for alkalinity.

Generally, more negative pKa value is a stronger acid, and more positive value of pKa is weaker acidity, or stronger alkalinity (base). The exact values of an acid must be determined experimentally. It cannot be calculated from the negativity of each element in a molecule, because it depends on the molecular structure and its functional group. We shall discuss the following example:

$$NH3 \quad + \quad H2CO3 \longleftrightarrow NH4 \quad + \quad HCO3$$

| stronger | stronger | weaker | weaker |
| base | acid | acid | base |

The nitrogen atom in NH3 has 5 electrons in its outermost orbit, and wants only three hydrogen atoms to be stable. If the orbit is completed to 8, then the molecular NH3 is difficult to release its three hydrogen atoms.

Thus it is a strong base. In the case of H2CO3 which is shown below, the oxygen has two electrons bonded to the two dot- pair; one carbon and one hydrogen and both repel each other.

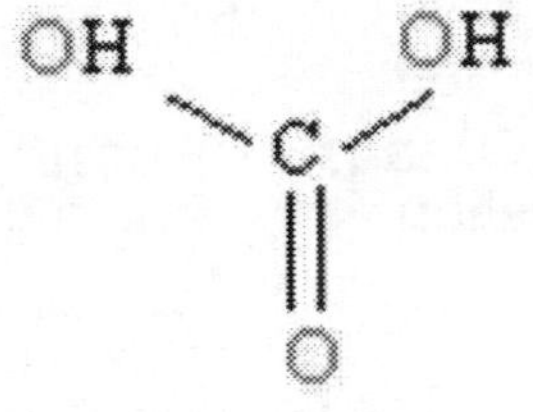

The force of repel to the hydrogen electron makes the oxygen to repel the hydrogen. Therefore, the molecular H2CO3 gives hydrogen to the NH3.

The pKa of the carbonic acid (H2CO3) is 6.36 and the pKa for the ammonium (NH3) is 9.24.

What is the series of acetylene?

Acetylene series is group of aliphatic (carbon atoms joined in a string open chain) hydrocarbons, each containing at least one triple carbon bond. The group resembles acetylene and has the formula of C_nH_{2n+2}, with acetylene being the simplest formula. There are mainly four groups of acetylene emerges from the chain. Generally, hydrocarbons are compounds that only contain H and C atoms of the formula C_nH_m, but they can be subdivided into four main groups, as shown in Figure (31).

Figure (31): Groups of Acetylene series

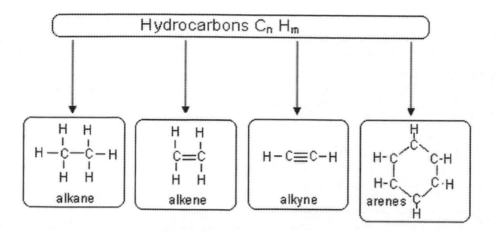

1- Alkanes

The general formula of alkane is C_nH_{2n+2}. The simplest is methane which is CH_4. Here are the first four groups of alkane.

Hydrocarbons which contain only single bonds are called alkanes. They are called saturated hydrocarbons because there is hydrogen in every possible location. This gives them a general formula C_nH_{2n+2}.

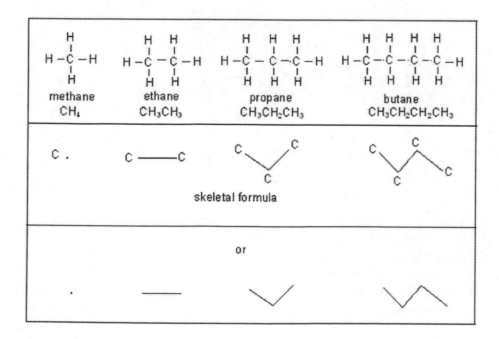

Methane can be added to ethane, propane, and butane to form methyl ethane, methyl propane, and methylbutane as follows

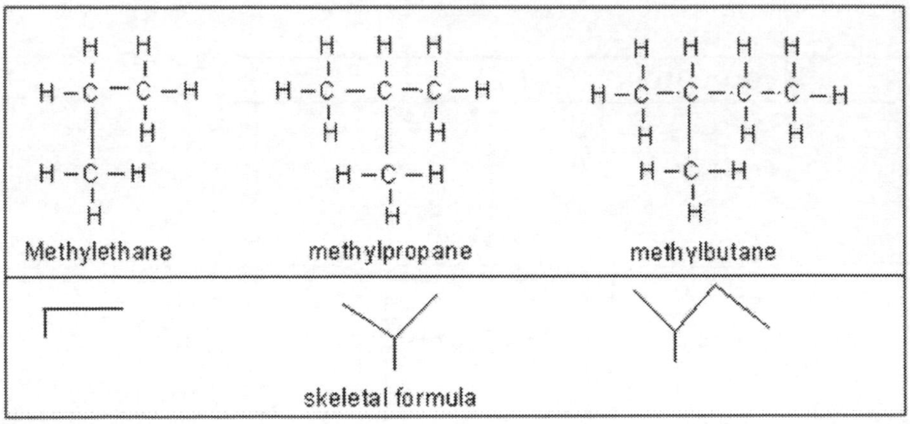

Methylethane   methylpropane   methylbutane

skeletal formula

Let's take the aliphatic heptanes, 2-chloroheptane, and 3-chloroheptane configurations:

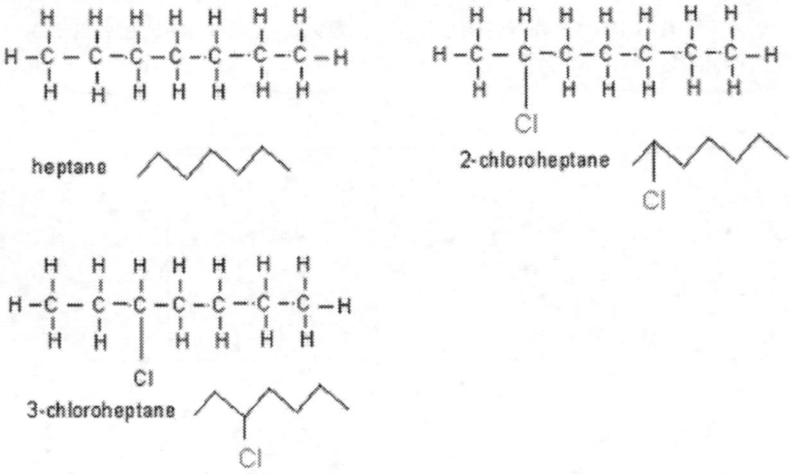

heptane

2-chloroheptane

3-chloroheptane

The suffix - ane is associated with the four groups of alkane which are gases, and prefixes penta, hexa, hept, oct, non, and dec are used for groups 5, 6, 7, 8, 9, and 10 are liquids. Liquids are up to $C_{17}H_{36}$. Alkanes are highly combustible clean fuels, forming heat, water, and carbon dioxide. Gasoline is a mixture of alkanes of C_5 to alkanes of C_{10}. Alkanes of C_{18} and above are solid at room temperature and found in petroleum jelly, paraffin wax, motor oils and lubricants. Asphalt is of a very high number of carbons.

2- Alkenes

Alkenes are one type of unsaturated hydrocarbons, because carbon atoms are with one or more double bonds, and therefore are holding fewer hydrogen atoms than they would if the bond was a single bond. Alkenes take the formula of C_nH_{2n}. Here are some alkenes:

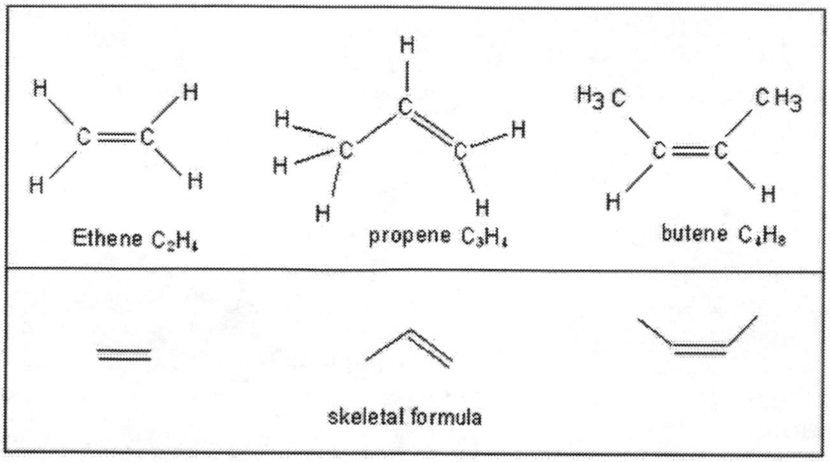

When alkenes have more than three carbons, they are isomers. This means that there are two or more different structural formulas that can be drawn for each formula. For example, the molecule of butylene has the following different structures:

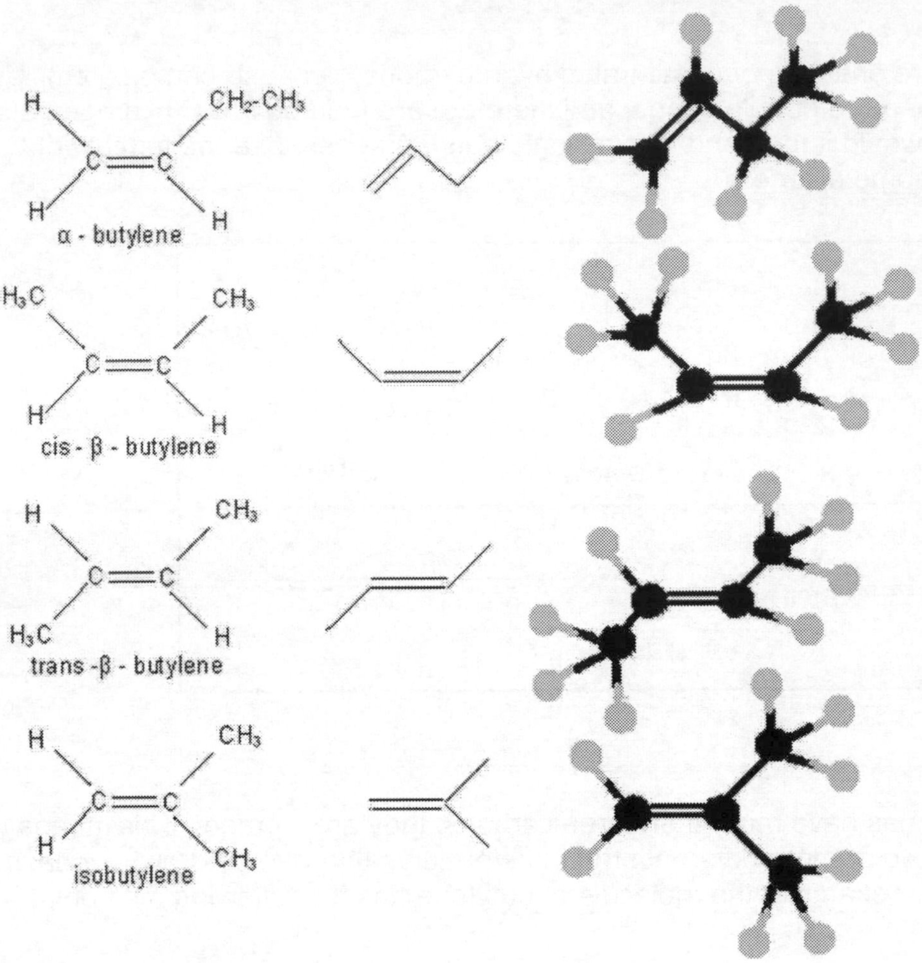

α - butylene

cis - β - butylene

trans - β - butylene

isobutylene

The double bonds between carbon atoms, which are relatively weak, alkenes, therefore need stronger acids to the double bonds of the carbons such as HCl, HBr, H2SO4, HI, etc. The majority of these reactions are exothermic, i.e. the output heat is higher than the combined heat of the reactants as shown below:

$$\text{Exothermic energy} = (99 + 63.5) - (63 + 72.5) = 27 \text{ kcal/mol}$$

3- Alkynes

Alkynes are hydrocarbons that have the formula of C_nH_{2n-2}. It has at least three bonds between two carbons. Traditionally, alkynes are known as acetylenes or acetylene series, and the simplest acetylene is the ethyne C2H2 which has the formula and shape of:

We shall talk about sigma bond and pi bond later.

Alkynes tend to be more electropositive than alkenes, because it has two hydrogen atoms and three bonds which make the space to be attached larger than the space of alkenes. The two hydrogen atoms are less positivity than four hydrogen atoms (hydrogen is a proton). Alkynes therefore tend to bind more tightly to a transition metals (columns 3 -12 in the periodic table and all of them end with I or 2 electrons in the outermost orbits) than alkenes. Triple bond is unstable and so alkynes are quite reactive. Since they have less hydrogen than alkenes, they are still able to act as acids, highly volatile, and combust readily such as propyne which is being used as a rocket fuel. Let's take the reaction of alkyne with a transition metal that ends with two electrons in its outermost orbit.

Figure (32) shows the reaction with a metallic base. The alkyne converts to alkene and then to alkane.

Figure (32): Alkane is produced from alkyne in two steps with the use of a transition metal base

In general, alkyne with transitional metal reaction is resonating to either alkene or alkane, as shown in Figure (33), http://www.ilpi.com/organomet/alkyne.html

Figure (33): Transitional reaction

4- Arenes

Arenes are aromatic hydrocarbons (compounds based on benzene rings) such as benzene and methylbenzene. Arenes are aromatic hydrocarbons, i.e. pleasant smells. They are based on benzene ring which has the simplest form of C_6H_6. The next simplest is methylbenzene (old name: toluene) which has the form $C_6H_5CH_3$. Benzene has two forms of structures; the form Kekulé and the new model form. In Kekulé form (the old one) the carbons are arranged in a hexagon, and bonds are alternating double and single between carbons. Kekulé form looks like ethene as it has double bonds between alternative carbons and one may think that benzene has reactions like ethene. Benzene is usually undergoes substitution reactions in which one of the hydrogen atoms is substituted by another atom from another substance. Benzene is a combination of 3 ethenes, and the difference between benzene and ethane is that ethane undergoes electrophilic carbon reaction and benzene undergoes substitution reaction, Figure (34).

Figure (34): Ethene with addition reaction and benzene with replacement reaction

$$CH2 = CH2 + HBr \longrightarrow CH3\overset{+}{C}H2 \longrightarrow CH3CH2Br + H$$

Benzene or benzol is a known carcinogen, though it is used as an additive in gasoline (now is limited), and as an important solvent and precursor in pharmaceutical drugs, plastics, rubber, and dyes. It is a natural component of crude oil and smoking and may be synthesized from other compounds. Breathing benzene can cause dizziness, drowsiness and unconsciousness. Long - term benzene exposure could cause effects on the bone marrow and cause anemia and leukemia. Long term exposure to high levels of benzene in the air can cause myelogenous leukemia, which is a cancer of the blood forming organs. Benzene has been determined by the Department of Health and Human Services (DHSHS), and the International Agency for Research on Cancer (IARC), and the Environmental protection Agency (EPA)as a known carcinogen.

What is IUPAC Nomenclature?

The International Union of Pure and Applied Chemistry (IUPAC) has developed a set of rules for giving a unique name for each organic compound. For example, alkanes come in all shapes and sizes. The simplest alkane hase all of its carbons chained together in a series (row). Here is a row of five carbons:

$CH_3 - CH_2 - CH_2 - CH_2 - CH_3$

Pentane has two parts; pent means five, and ane means the compound alkane. The table below shows the appropriate base part for other number of carbons.

Number of carbons	Base	Number of carbons	Base
1	Meth	11	Undec
2	Eth	12	Dodec
3	Prop	13	Tridec
4	But	14	Tetradec
5	Pent	15	Pentadec
6	Hex	16	Hexadec
7	Hept	17	Heptadec
8	Oct	18	Octadec
9	Non	19	Nonadec
10	Dec	20	Icos

Now you know the twenty names of alkane bases, then you can name other number of carbons

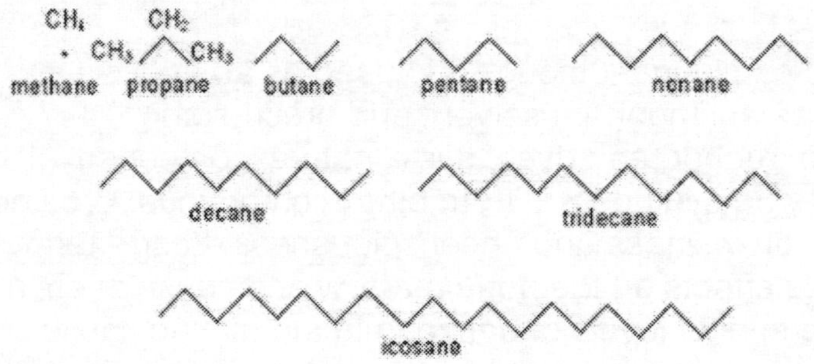

methane   propane   butane   pentane   nonane

decane   tridecane

icosane

One can see that alkane can still have the same number of carbons even the chain is not on the same horizontal line as shown with tridecane above. Horizontal lines can also be represented in rings which do not have CH3 anywhere if the molecule is stable as figured below:

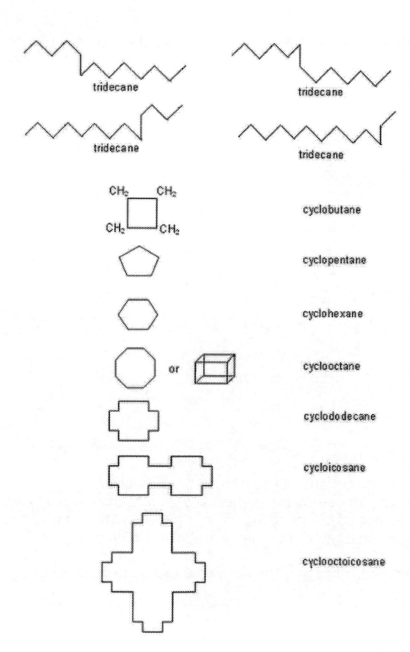

tridecane

tridecane

tridecane

tridecane

CH₂   CH₂

CH₂   CH₂

cyclobutane

cyclopentane

cyclohexane

or

cyclooctane

cyclododecane

cycloicosane

cyclooctoicosane

Numbering of IUPAC (International Union of Pure and Applied Chemistry)

Numbering can start from left or right depends on the location of other carbon atoms attached to the chain as indicated in Figure (35).

Figure (35): Correct and incorrect numbering of molecules

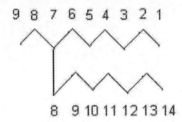

This is correctly numbered because it is the longest chain

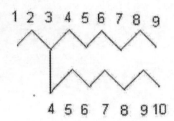

This is not correctly numbered because it is not the longest chain

Figure (38) is again for nonane because it has nine carbons, but has an attachment of CH_3. The base for one carbon is *meth* and the attachment is defined by IUPAC as *yl,* so the connected branch to the chain is called *methyl,* so we have a methylnonane. We show three types of connection *yl* as in Figure (36).

Groups can also be attached to rings and rings can be attached to rings. See the following groups:

Figure (36): Three types of connections to the chain nonane

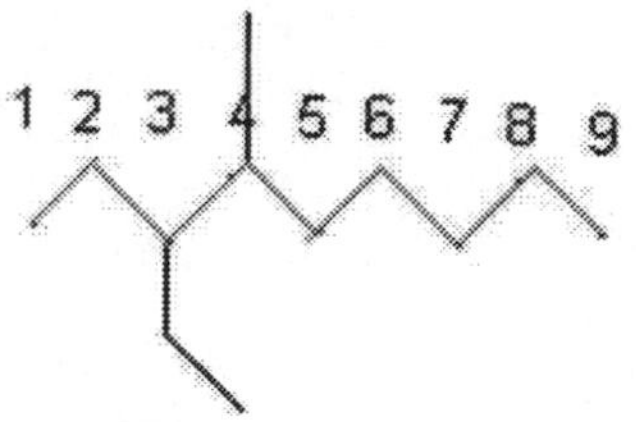

3-ethyl-4-methylnonane
correct

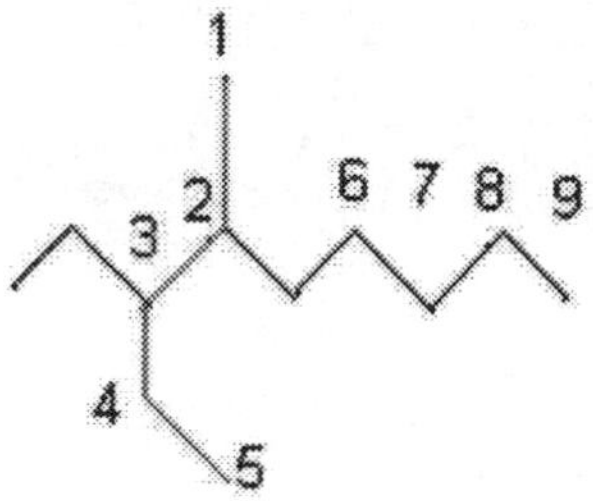

2-methyl-3-ethylpentane
incorrect

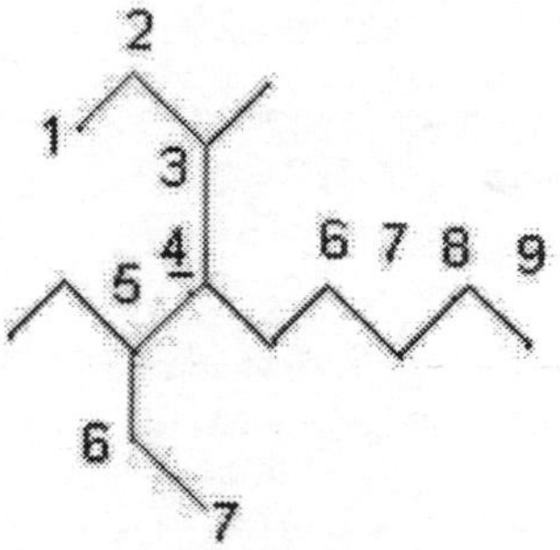

3-ethyl-methylethylnonane   correct
3-methyle-5-ethyleheptane   incorrect

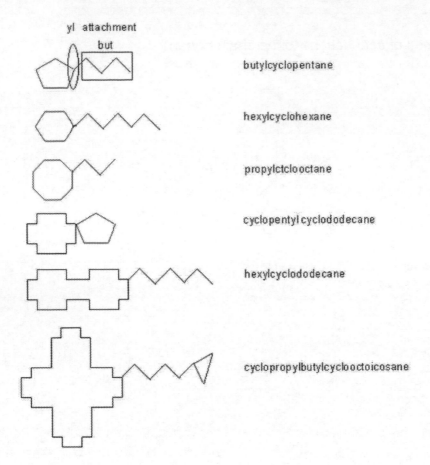

yl attachment

but

butylcyclopentane

hexylcyclohexane

propylctclooctane

cyclopentyl cyclododecane

hexylcyclododecane

cyclopropylbutylcyclooctoicosane

Let's consider the following two structures of the same number of carbons and hydrogens. The one on the left the two carbons are attached to the middle carbon which is attached to the cyclohexane. The one on the right the three carbons are attached in series, and the end of the chain (CH_2) is attached to the cyclohexane. IUPAC names the nodal connection as iso and the series as a straight name.

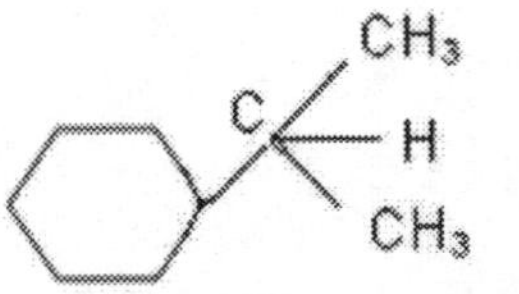

CH_3

C — H

CH_3

CH_2 CH_3

CH_2

isopropylcyclohexane

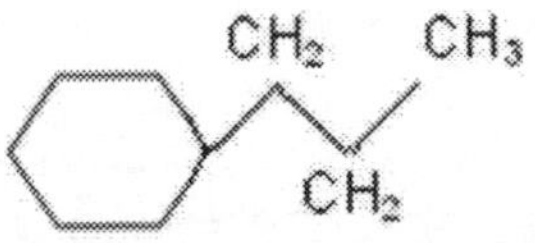

propylcyclohexane

Above nomenclature is not difficult if we use a step by step method for numbers and names. We shall treat some other names which are called systematic names which gives numbers for locations of attached molecules. One has to remember that nodal connection is not stable molecule as the straight chain-connection, so

64

the molecule with more nodal connections is more volatile and has lower boiling point than molecules with less nodal connection, providing that both have the same number of carbons and hydrogens. Figure (37) has more complicated names.

Figure (37): Common names and systematic names

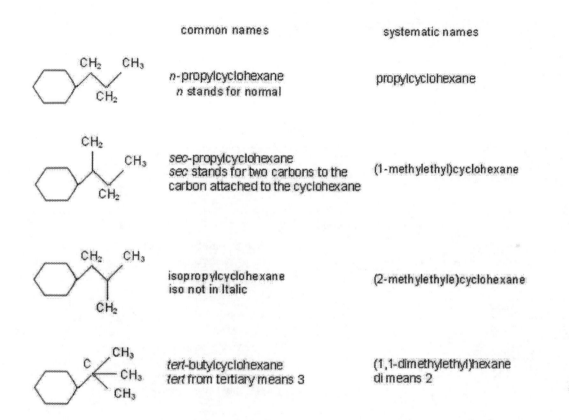

If the side chain is more than 4 carbons, it is commonly named systematically. Organic compounds can have more than one group attached to the longest chain. The following table has names for simple groups prefix and complex group prefix.

Number of groups	Simple group prefix	Complex group prefix
2	di	bis
3	tri	tris
4	tetra	tetrakis
5	penta	pentakis
6	hexa	hexakis
7	hepta	heptakis

Let's take some examples as shown in Figure (38).

Figure (38): More than two simple groups on the main chain

In Figure (38), the top configuration has the numbering from right to left, and the methyl hits carbon at number 3. Numbering can not be from left to right as the methyl hits the main chain at carbon number 6. In configuration number two from the top, numbering should also be from left to right for the same reason as explained above. In the third configuration, numbering is correct from right to left as the methyl hits the carbons at 3,4,and 6, not at 3,6, and 8. The fourth configuration can be from either sides because the group propyl or (methylethyl) hits the main chain at 3,6,8 or 4,6,9 or (3,6,8 or 4,6,9) respectively, the difference between the hitting points are the same.

Consider connections of rings, Figure (39). In rings connection, there is no start or end where we can number the connections. We start numbering at the carbon that one of the groups is connected to. Then numbering is done in the direction of the closest group. In Figure (39), the cyclohexane is connected to three groups of methane. So, the nomenclature as per the IUPAC is 1,2,4-trimethylcyclohexane.

Therefore, the green numbering is wrong. Note that three connections are supposed to be horizontal to ring. If the connections are inward to the paper plain, then the name of the component is preceded with cis-, and if the connection is outward, it is preceded with trans-. Both cis and trans are in Latin.

Figure (39): Numbering prefixes of rings with connected groups

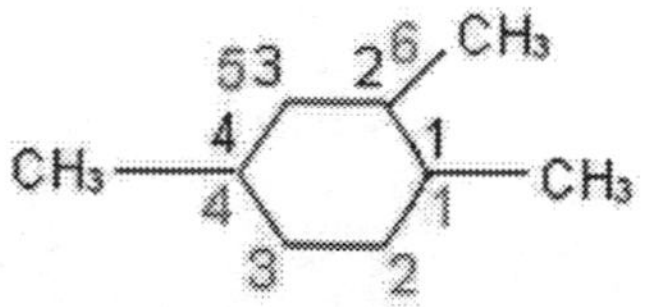

1,2,4-trimethylcyclohexane

1,2,3,4 numbering is correct
1,2,3,4 numbering is wrong

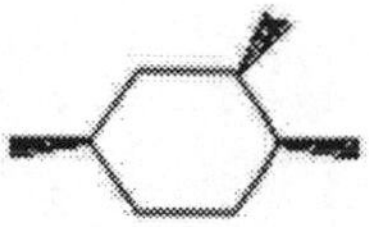

cis-1,2,4-triethylcyclohexane

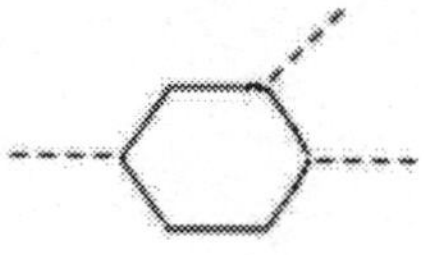

trans-1,2,4- triethylcyclohexane

Now, if you understand the previous demonstration of the IUPAC nomenclature, you can name hundreds of groups. However, there are still many complicated molecules needed to explain so that one can understand the whole subject, Figure (40).

Figure (40): More complicated chains and rings

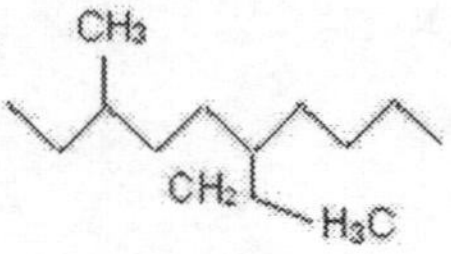

The chain is a decane.
The chain has a methane and an ethyl group.
We number from the left because the methyl is closer to the end of the chain. The groups will be arranged alphabetically. So ethane is before Methane. Therefore, the correct name is 6-ethyl-3-methyldecane.

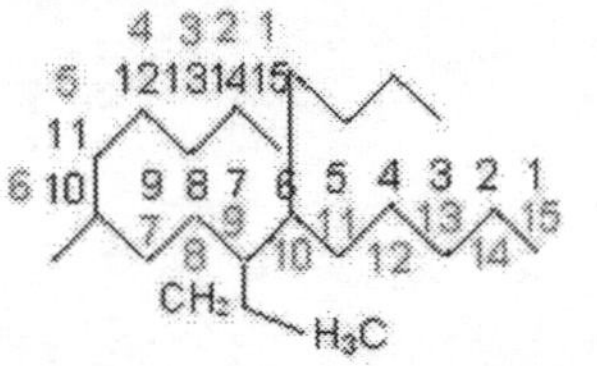

The chain is a pentadecane. The groups attached to alphabetically are butyl, ethyl, and methyl. The locations of the groups are: butyl is at carbon 6 (we start from right to left because the group butyl is the closest to the end of the chain.), ethyl at 7, and methyl at 10. Since we can also start numbering from the top of the chain as colored in green, we find the group methyl is also at carbon 6. Therefore, ethyl at 9, butyl at 10, and methyl at 6. So the molecule has two names:
6-butyl-7-ethyl-10-methylpentadecane, or
10-butyl-9-ethyl-6-methylpentadecane

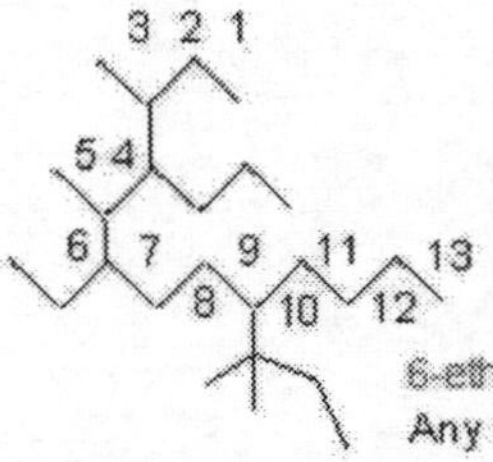

6-ethyl-3-methyl-5-methyl-9-(1,1,1dimethylmethylethyl)tridecane
Any other name? yes

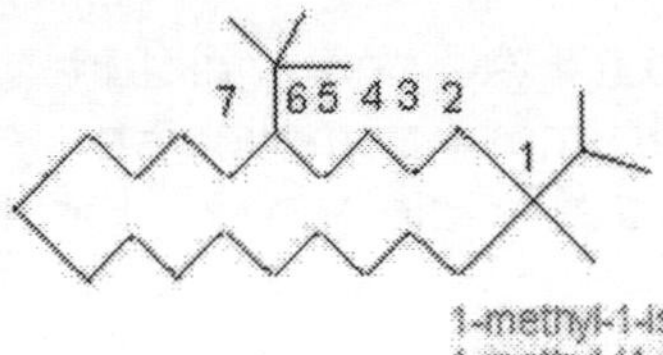

1-methyl-1-isopopyl-6-isobutylcycloicosane or
1-methyl-(1-methylethyle)-6-(1-ethyl-1-isoethyl)cycloicosane or
1-methyl-(1-methylethyle)-6-(1-ethyl-1methyl-1-methyl)cycloicosane.

What are functional groups?

An atom or group of atoms, that replaces hydrogen in an organic compound and that defines the structure of a family of compounds and determines the properties of the family, http://www.answers.com/topic/functional-group

Alkanes, alkenes, alkynes, and arenes are functional groups. In addition, there are about 30 or so functional groups that determine the properties and reaction chemistry of molecules. It is essential that professional chemists, to be able to name organic molecules, predict solubility, chemical reactivity, and the spectra of drug effectiveness to recognize the functional groups. For example, drug morphine may have several functional groups as shown in Figure (41), http://www.chemistry-drills.com/functional-groups.php?=simple

Figure (41): Morphine drug with several functional groups

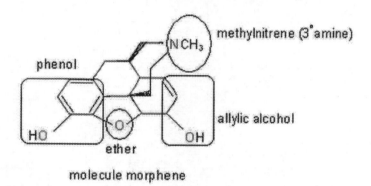

The table below lists functional groups of interest to understand the aim of this book.

Class of compound	Functional group	IUPAC name	formula
halide	-F fluoro -Cl chloro -Br bromo -I iodo	R-X R represents alkyl, X represents halogen 2-chloropropane	$CH_3CHClCH_3$
primary alcohol	-OH	R=CH_3CH_2 1-propanol	RCH_2OH

Secondary alcohol	-OH	R_2= 2(CH_3CH_2) 1-diethylisopentane	(structure) R—C—OH with R above and R_2 below R_2CHOH
Tertiary alcohol	-OH	R_3=3(CH3CH2) 1-triethyleisoheptane	(structure) R—C—OH with R above and R below R_3COH
ether	-O-	ethylmethyether or methylethylether	(structure) —C—O—C—C— $CH_3OCH_2CH_3$
aldehyde	(structure) —C—H with =O	propanal	(structure) —C—C—C—H with =O CH3CH2CHO
ketone	(structure) —C— with =O	2-pentanone	(structure) C—C—C—C— with =O $CH_3COCH_2CH_2CH_3$
organic acid	(structure) —C—OH with =O	propanoic acid or ethanecarboxylic acid	(structure) C—C—C—C— with =O $CH_3COCH_2CH_2CH_3$
ester	(structure) —C—O— with =O	double bond oxygen ⌐ methylpropanoate single bond oxygen ⌐	(structure) —C—C—C—O—C— with =O CH_3CH_2 $COOCH_3$
amine	(structure) —N—	1-propanamine	(structure) —C—C—C—N— CH3CH2CH2NH2
amide	(structure) —C—NH with =O	propanamide	(structure) —C—C—C—N— with =O $CH_3CH_2CONH_2$
acid chloride	(structure) —C—Cl— with =O	propanoyl chloride oyl replace e of propane when connected to oxygen O	(structure) —C—C—C—C—Cl with =O $CH_3CH_2CH2OCl$

acid anhydride	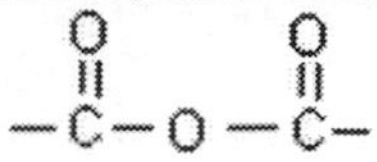 (structure: $-C(=O)-O-C(=O)-$)	ethanoic anhydride	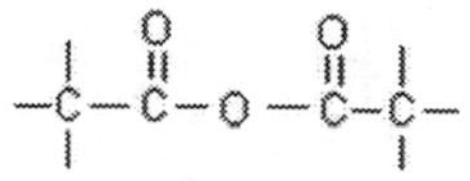 (structure) $CH_3COOCOCH_3$
nitrite	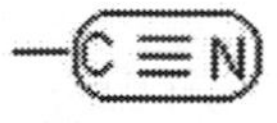 (structure: $-C\equiv N$) cyano group	Ethanenitrile or acetonitrile	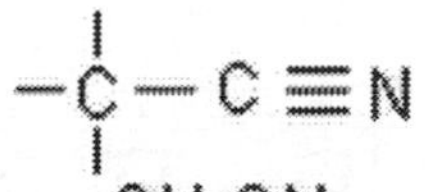 (structure) CH_3CN
Amino acid alanine	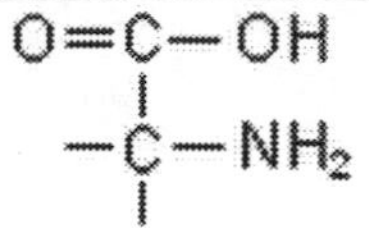 (structure: $O=C-OH$, $-C-NH_2$)	2-aminopropanoic acid	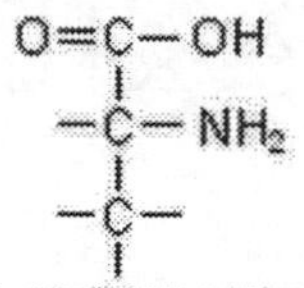 (structure) $CH_3CH(NH_2)COOH$
Amino acid isoleucine		2-amino-3-methylpentanoic acid	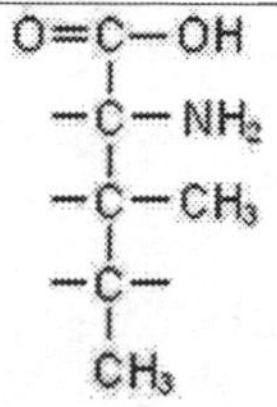 (structure) $CH(NH_2)CH(CH_3)CH_2CH_3COOH$
tyrosine		2-amino-3-(4-hydroxyphenyl)-propanoic acid	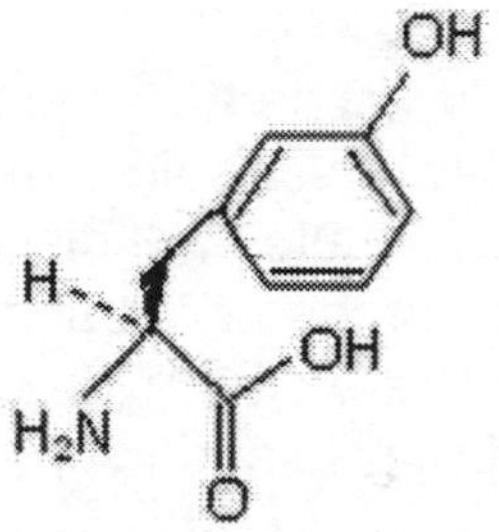 (structure) $OHC_6H_4C_2CHH_2NCOOH$
trans alkene	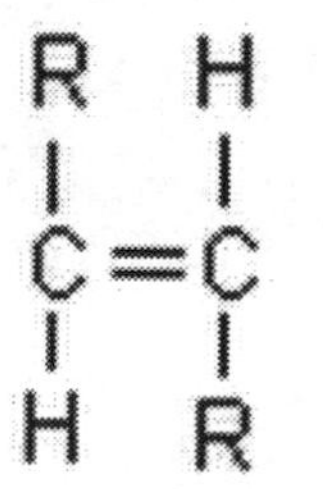 (structure) $R_2C_2H_2$	alkenyle lithium	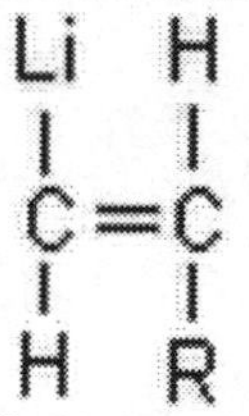 (structure) RC_2H_2Li
trans-alkene		trans-3-methylhex-3-ene	 (structure) C_6H_{11}

cis alkene	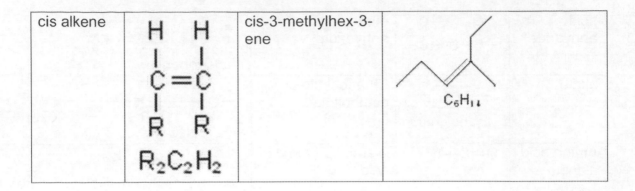	cis-3-methylhex-3-ene	

What is solubility?

1-21 Solubility

Solubility will be affected by the polarity of both the solute and the solvent. Generally, polar solute molecules will dissolve in polar solvents and non-polar solute molecules will dissolve in non-polar solvents. For example, water is polar because of the unequal sharing of electrons, methane is non-polar because the carbon shares the hydrogen atoms uniformly, and phospholipids are surfactant, i.e. hydrophilic and lipophilic, Figure (42).

Figure (42): Examples of polar and non-polar molecules

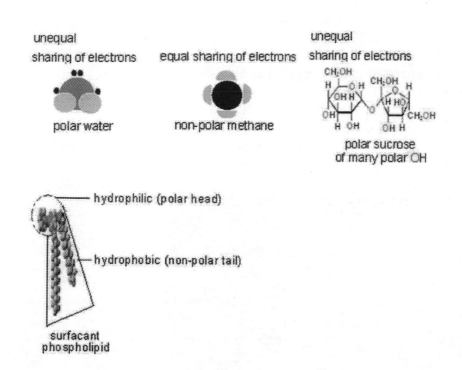

The unequal sharing of electrons within a molecule leads to the formation of two different poles (dipole); one negative and one positive.

Atoms on the right four groups of the periodic table are generally electronegative and exert a greater pull on electrons of the left groups, For example, oxygen exert greater pull on hydrogen than carbon, and fluorine exerts greater pull on hydrogen more that oxygen, and so on. Figure (43) shows the hydrogen is closer to fluorine than oxygen, because the latter has less negativity than fluorine.

Figure (43): Hydrogen is closer to fluorine than oxygen

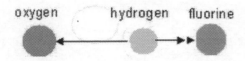

fluorine exerts greater pull on hydrogen than oxygen

In the polar solvent molecule the positive ends will attract the negative ends of solute molecule. Figure (44) shows water (polar) and boron difluoride Bf_2 (polar) and their attraction.

Figure (44): Attraction between positive and negative ions

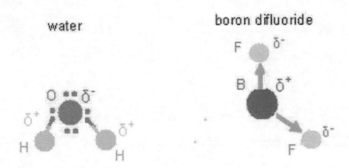

fluoride attracts hydrogen and oxygen attracts boron

Are the protons and neutrons the smallest particle in an atom?

No. Scientists have proved theoretically and experimentally that the protons and neutrons are made up of even smaller particles, called quarks. Particles that cannot be broken further such as quarks are sometimes called fundamental particles.

Scientists now believe that the nucleus of an atom (nucleus has only protons and neutrons) has protons and neutrons made of smaller particles: quarks and three other types of particles—leptons, force-carrying bosons, and the Higgs boson—which are truly fundamental and cannot be split into anything smaller. Higgs bosons have not yet been proven experimentally. In the 1960s American physicists Steven Weinberg and Sheldon Glashow and Pakistani physicist Abdus Salam (they shared the Noble prize for their discovery), developed a mathematical description of the nature and behavior of elementary particles. The term elementary particles has the same meaning as fundamental particles but is used more loosely to include some subatomic particles that are composed of other particles. Their theory, known as the Standard Model of Particle Physics, has greatly advanced the understanding of the fundamental particles and forces in the universe. Some questions about particles, including Boson particles, remain unanswered by the standard model, and physicists continue to develop a theory that will explain even more about many particles emitted from the universe.

Atoms can be classified into one of the two categories called Fermions or Bosons. Fermions are fundamental particles forming protons and neutrons. Fundamental bosons carry forces between particles and give particles mass. Boson is the name for a generic class of particles. The Higgs boson is one (if it exists) but so are many other particles. All the particles which carry forces (gluons, the W and the Z and the photon, also the graviton, if there is one.) are bosons. Quarks, electrons and neutrinos, on the other hand, are fermions.

The difference between them is just spin. But in this context, spin is a quantum number of angular momentum. It is a bit like the particle is spinning, but that is really just an analogy, since point-like fundamental particles could not spin, and anyway fermions have a spin such that in a classical analogy they would have to go round twice to get back where they started. Quantum mechanics is full of semi-misleading analogies like this. Regardless, spin is important.

Bosons have, by definition, integer spin. The Higgs has zero, the gluon, photon, W and Z all have one, and the graviton is postulated to have two units of spin. Quarks, electrons and neutrinos are fermions, and all have a half unit of spin. The table below shows the difference.

| Fermions | Half-integral spin | Only one per state | Electrons, protons, neutrons, quarks, neutrinos |
| Bosons | Integral spin | Many can occupy the same state | Photons, 4He atoms, gluons, gravitons |

It was proven by Wolfgang Pauli, an Austrian American, that no two electrons have the same momentum and location. This was called the Exclusion Principle. The Exclusion Principle was developed to include all particles that obey such a principle. Fermions, in honour of the Italian American physicist Enrico Fermi, which include quarks and leptons, obey the Exclusion Principle theory.

German American physicist and an Indian mathematician Satyendra Bose proved that Bosons, see Figure (45) below, suggested that they did not obey the Exclusion Principle. Bosons, in honor of Bose, include photons, gluons, and weak forces. Higgs bosons were proven theoretically, but not yet experimentally.

The Exclusion Principle can be based on the number of fermions. If the number of fermions is even, then the atom does not obey the Exclusion Principle. If it is odd, then it obeys the Exclusion Principle.

Example: Hydrogen has one proton (one proton has three quarks), and one electron (one electron is one lepton); therefore, it does not obey the exclusion theory. An atom of heavy hydrogen (deuteron) has one proton (3 quarks), one neutron (3 quarks), and one electron (1 lepton). Therefore, the number of quarks is odd, and it obeys the Exclusion Principle. It concludes that a deuteron cannot have the same properties as another deuteron atom. On the other hand, properties of the hydrogen atoms can be identical to the properties of another hydrogen atom.

Figure (45): Fermions and bosons forming the atom

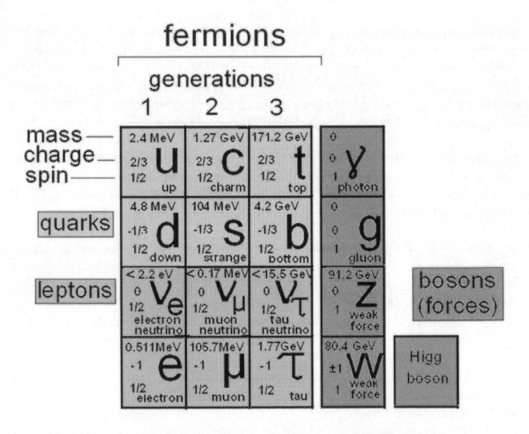

What does the Exclusion Principle mean in practical life?

Consider copper and iron wires for conducting electricity. Electrons in the copper wire follow the Exclusion Principle, whereas electrons in the iron wire only slightly follow the Exclusion Principle. Thus, copper wires are better than iron copper in conducting electricity. Laser and photons (light) do not obey the exclusion principle; they are bosons, and have identical properties. This characteristic of light and laser makes them form consistent and solid beams that can travel a long distance.

Bosons are similar to the gravity of our earth. Gravity cannot be seen but it carries a stone, for example, from a higher level to the ground level. Bosons carry the four basic forces in the universe: the electromagnetic, the gravitational, the strong (gluons which hold the quarks together), and the weak forces that cause the atom to decay (see beta decay).

What are the functions of nuclear forces?

The electromagnetic force binds electrons to atomic nuclei (clusters of protons and neutrons) to form atoms.
The gravitational force acts between massive objects. Although it plays no role at the microscopic level (between atoms), it is the dominant force in our everyday life and throughout the universe.
The strong force (gluons) is responsible for quarks "sticking" together to form protons, neutrons and related particles.
The weak force facilitates the decay of heavy particles into smaller siblings.

What is The Standard Model?

The most fundamental building block of all matter – the matter that makes up every thing from prokaryotes and eukaryotes to people to galaxies to supernova, and cannot be broken down to any thing smaller is particles known as subatomic particles. For example, the atom is made of protons, neutrons, and electrons. Protons and neutrons are made of particles called quarks, and electrons are made of leptons. Figure (46) shows a helium atom with its subatomic quarks and leptons.

Figure (46): Subatomic quarks of helium atom

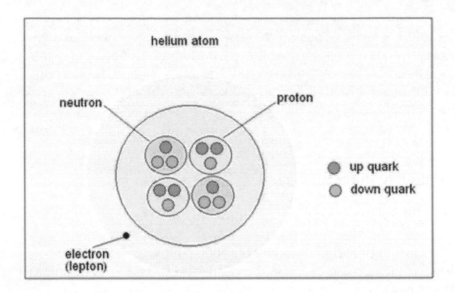

What is a force? Let's take this example. You may have heard of gravity. Gravity is the force that all objects with mass exert upon one another, pulling the objects closer together. It causes a ball thrown into the air to fall to the earth, and the planets to orbit the sun.

The tiny particles that make up matter, such as atoms and subatomic particles, also exert forces on one another. These forces are not gravity, but are Special Forces that only these particles use.

There are several kinds of forces that particles can exert on one another. These forces can cause one particle to attract, repel, or even destroy another particle. For example, one kind of subatomic force, known as the strong force, binds quarks together to make protons, neutrons, and other particles.

Can you define particles of antimatters?

Particles of Antimatter

Paul Dirac, a British physicist proposed a theory of antiparticles that combine to form antimatter. Antiparticles and particles have the same mass, but their electric charge and colour charge are different. The electric charge and colour charge determine how particles react with each other. Both Fermions and bosons have their own antiparticles.

As protons consist of quarks, antiprotons consist of antiquarks; one antiproton has two up antiquarks and one down antiquark. Similarly, one anti neutron has two down antiquarks and one up antiquark. The antielectron is called a positron, and the muon and the tau have their counterpart's antimuon and antitau. The antiparticles of neutrinos are called antineutrino. Neutrinos and antineutrinos have no electric charge or colour charge. The Antineutrino accompanies the lepton when a proton and neutron decay, and the antineutrino and neutrino balance the output of the decay, Figure (47). Reaction that absorbs neutrino does not absorb antineutrino and vice versa.

Figure (47): Decays of proton, neutron, and fermions

$$n \longrightarrow p + w^- = p + e^- + \bar{\nu}_\mu \qquad p \longrightarrow n + w^+ = n + e^+ + \nu_\mu$$

$\bar{u} + d = -2/3 - 1/3 = -1$
$w^- = -1$
$\mu^- + \bar{\nu}_\mu = +1$

$u + \bar{d} = +2/3 + 1/3 = +1$
$w^+ = +1$
$\mu^+ + \nu_\mu = -1$

Question – can you draw the outcome of an up quark and a down antiquark in terms of electrons and electron neutrinos? Can you determine the outcome of the down quark and up antiquark in terms of positrons and electron antineutrinos?

Define subshells and orbitals

The table below shows letters to the different subshells. It also tells us how many orbitals are in each subshell and the shape of the orbitals in that subshell.

Subshells		
Subshell	# of Orbitals	Orbital Shape
s	1	spherical
p	3	dumbbell
d	5	dumbbell
f	7	dumbbell

Figure (48) shows the subshells and their arrangements.

Figure (48): Subshells with number of orbits

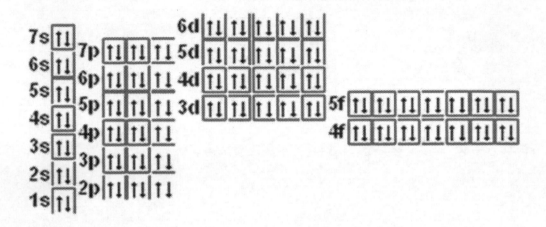

Question: How many electrons in "p" subshell?

Answer: 6 x 2 = 6

Each orbital or subshell holds no more than 2 electrons

Shell - - the shell represents the orbit around a nucleus. The first shell (or orbit) is close to the nucleus. The second shell is a little farther out from the nucleus. The energy of the orbits increase as we move away from the nucleus

Ok, now that we understand the concept of orbitals, subshells, and shells we can look at the distribution of electrons in specific atoms. Figure (49) shows the level of energy and the distribution of electrons.

Figure (49): Electronic distributions in orbitals

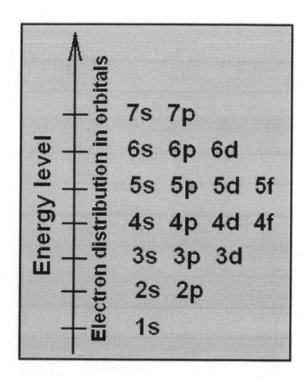

Examples:

Hydrogen (H): $1s^1$ subshell is not filled
Helium (He): $1s^2$ subshell is filled
Beryllium (be): $1s^2 2s^2$ subshell is filled
Lithium (Li): $1s^2 2s^1$ subshell is not filled
Nitrogen (N): $1s^2 2s^2 2p^3$ subshell is not filled
Neon (Ne): $1s^2 2s^2 2p^2 2p^2 2p^2$ subshell is full
Sodium (Na): $1s^2 2s^2 2p^2 2p^2 2p^2 3s^1$ subshell is not filled
Calcium (Ca): $1s^2 2s^2 2p^2 2p^2 2p^2 3s^2 3p^2 3p^2 3p^2 4s^2$ subshell is filled
Scandium (Sc): $1s^2 2s^2 2p^2 2p^2 2p^2 3s^2 3p^2 3p^2 3p^2 4s^2 3d^1$ subshell is not filled

Also, as you change Shells (or orbits) you are changing energy. Figure (50) shows you that energy increases as you get farther from the nucleus. An electron in the 3rd shell has more energy than an electron in the 1st shell.

In each Shell you see the various subshells. For example, in the 2nd Shell there are 2 subshells: an "s" subshell and a "p" subshell, Figure (50).

Figure (50): Subshells in orbits

m = 0 for s   l=0 for s   n=1 for s
m= 1,0,-1 for p   l=1 for p   n=2 for p
m= 2,1,0,-1,-2 for d   l=2 for d   n=3 for d
m= 3,2,1,0,-1,-2,-3 for f   l=3 for f   n=4 for f

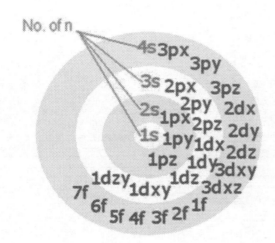

How electron shells are filled?

Aufbau Principle

The orbitals of an atom are filled from the lowest energy orbitals to the highest energy orbitals.

Orbitals with the lowest principal quantum number (n) have the lowest energy and will fill up first. Within a shell, there may be several orbitals with the same principal quantum number. In that case, more specific rules must be applied. For example, the three p orbitals of a given shell all occur at the same energy level. So, how are they filled up? ans: all the three p orbitals have same energy so while filling the p orbitals we can fill any one of the Px, Py or Pz first. it is a convention that we chose to fill Px first ,then Py and then Pz for our simplicity.

Hund's Rule

According to Hund's rule, orbitals of the same energy are each filled with one electron before filling any with a second. Also, these first electrons have the same spin.

This rule is sometimes called the "bus seating rule". As people load onto a bus, each person takes his own seat, sitting alone. Only after all the seats have been filled will people start doubling up.

Pauli Exclusion principle

No two electrons can have all four quantum numbers the same. What this translates to in terms of our picture of orbitals is that each orbital can only hold two electrons, one "spin up" (+½) and one "spin down" (-½).

Figure (51) shows the sequence of filling orbits

Figure (51): Sequence of filling s and p orbits

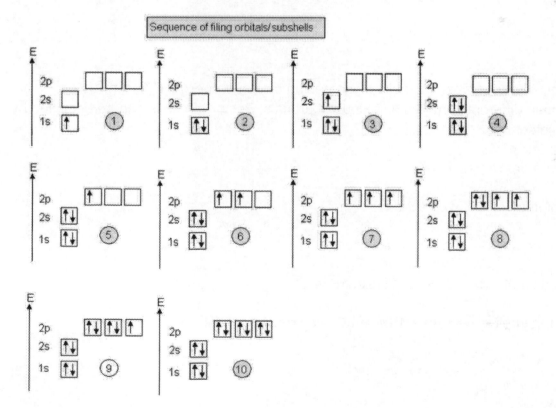

How many electrons in all orbitals?

As we said, there are 4 types of orbitals:

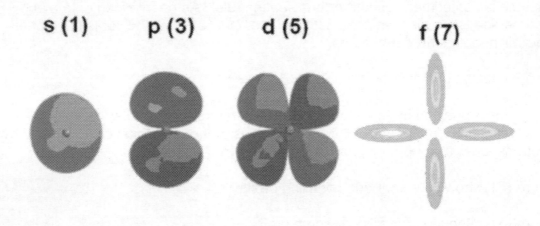

s (1)   p (3)   d (5)   f (7)

Orbitals are grouped in shells of increasing size and energy. Different shells contain different numbers and kinds of orbitals. Each orbitals is can be occupied by two electrons.

Figure below has 4 shells:

First shell – 1s (two electrons)

Second shell – 2s + 2p (8 electrons)

Third shell – 3s + 3p +3d (18 electrons)

Fourth shell – 4s + 4p + 4d + 4f (32 electrons)

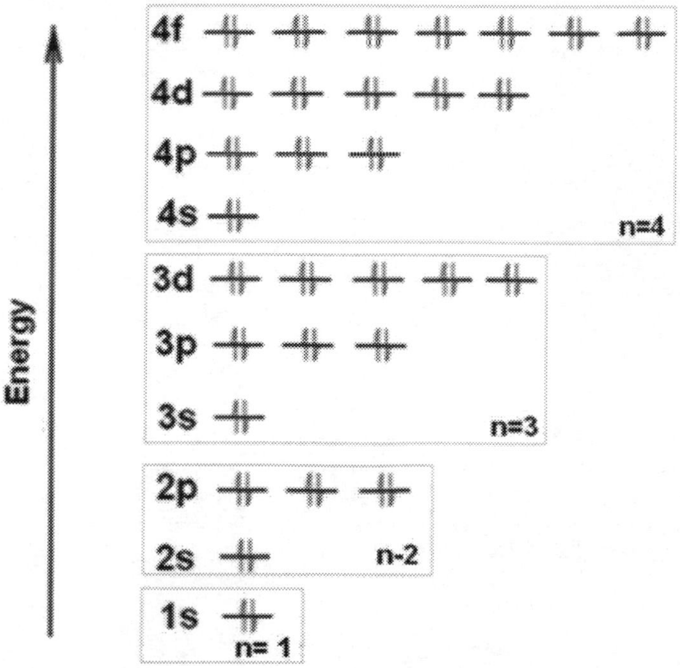
Energy
4f
4d
4p
4s
n=4
3d
3p
3s
n=3
2p
2s
n-2
1s
n= 1

Can you show orbitals on periodic table?

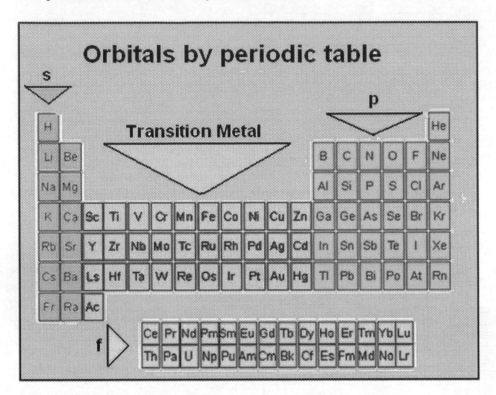

Can you draw covalent bonds with electron shells?

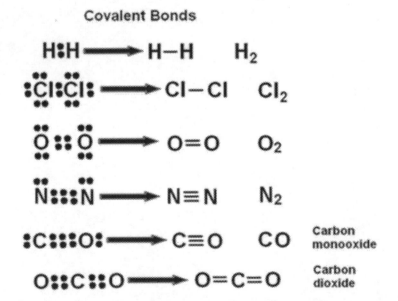

Can you show energy bond of two hydrogen atoms?

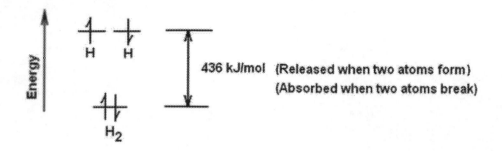

436 kJ/mol   (Released when two atoms form)
(Absorbed when two atoms break)

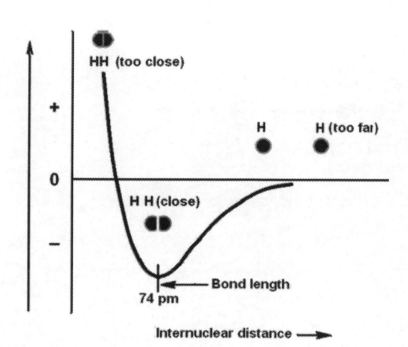

How are shapes of orbitals formed?

1- Sp^1

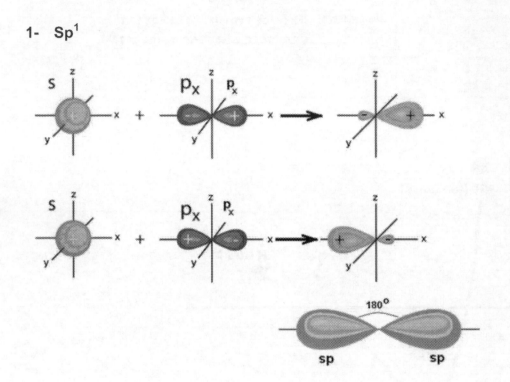

2- Sp²

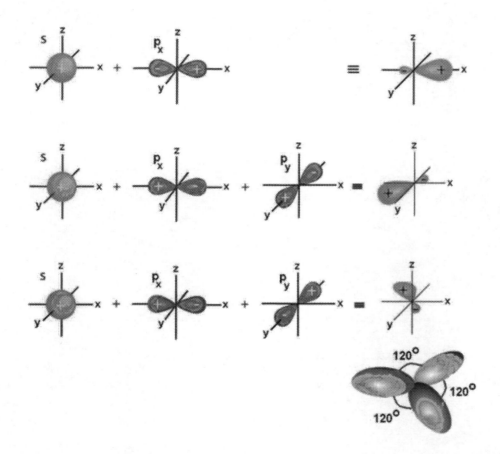

3- Sp³

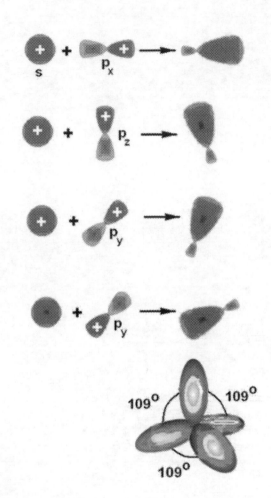

Can you show angles between orbitals?

Pure atomic orbitls of centarla atom	Hybridization of the central atom	Number of hybrid orbitals	Shape of hybrid orbitals
sp	sp	2	Linear
spp	sp²	3	Trigonal Planar
sppp	sp³	4	Tetrahedral
spppd	sp³d	5	Trigonal Bipyramidal
spppdd	sp³d²	6	Octahedral

Can you show the bond between metallic and non-metallic elements?

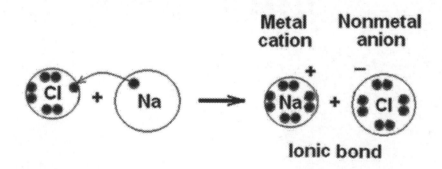

Metal cation   Nonmetal anion

Ionic bond

	Br^{1-}	O^{2-}	N^{3-}
Na^{1+}	NaBr	Na_2O	Na_3N
Mg^{2+}	$MgBr_2$	MgO	Mg_3N_2
Al^{3+}	$AlBr_3$	Al_2O_3	AlN
Fe^{3+}	$FeBr_3$	Fe_2O_3	FeN
Cu^{1+}	CuBr	Cu_2O	Cu_3N

Can you show covalent bonds using Lewis dot?

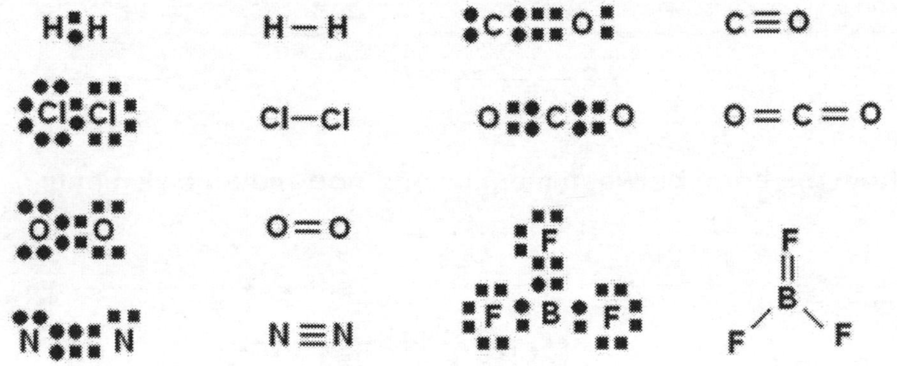

H—H

Cl—Cl

C≡O

O=C=O

O=O

N≡N

Can you show orbital shapes of all orbitals?

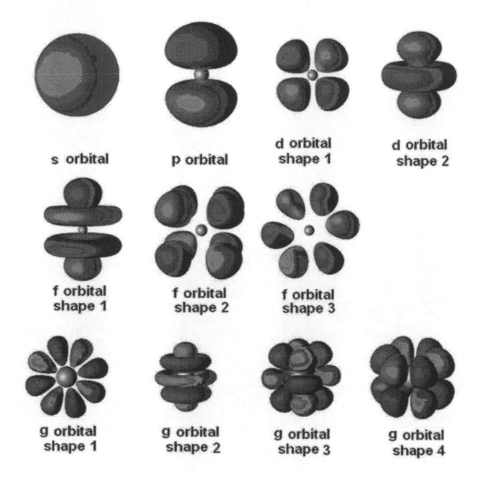

s orbital   p orbital   d orbital shape 1   d orbital shape 2

f orbital shape 1   f orbital shape 2   f orbital shape 3

g orbital shape 1   g orbital shape 2   g orbital shape 3   g orbital shape 4

Can you show orbitals 1s, 2s and 3s together with 2p?

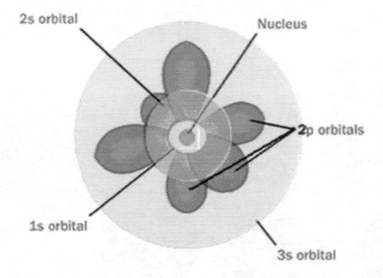

2s orbital

Nucleus

2p orbitals

1s orbital

3s orbital

Can you show orbitals s,p,d, and f?

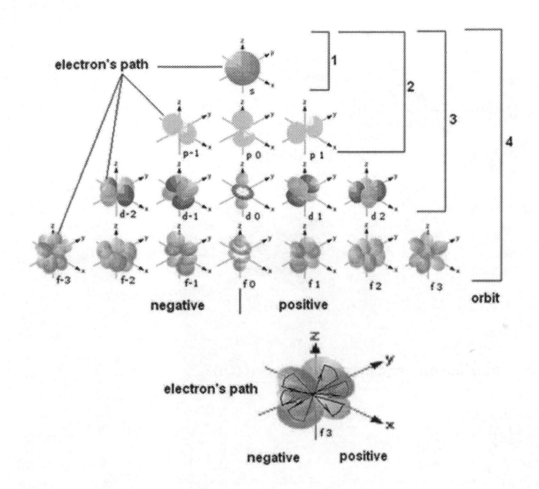

Can you show orbitals of ions of helium positive, hydrogen negative, oxygen negative, fluorine negative and calcium two positives?

Ion	Stable Electrons	Electrons in Ion	Electron configuration in Ion
He+	2	1	1s1
H-	1	2	1s2
O-	8	9	1s22s22px2py2pz1
F-	9	10	1s22s22p6
Ca2+	20	18	1s22s22p63s22p6

H-, F-, and Ca2+ are similar to noble gases in orbital configuration.

Can you draw the orbital Diagram for Hydrogen (H₂)?

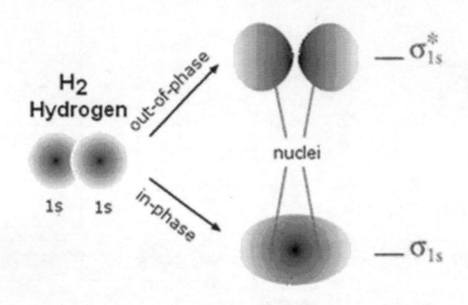

Can you draw the orbital diagram for oxygen (O₂)?

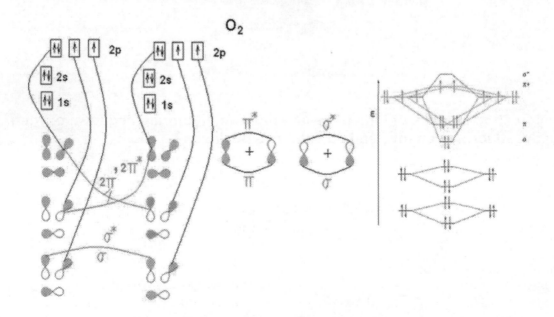

Can you draw the orbital diagram of water (H₂O)?

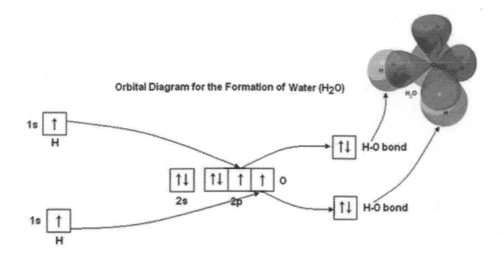

Orbital Diagram for the Formation of Water (H₂O)

What does hybridization mean?

The process of mixing atomic orbitals to form a set of new equivalent orbitals is termed as hybridization. There are three types of hybridization (They were explained earlier, but we shall deal with them in different ways):

(a) sp3 hybridization (involved in saturated organic compounds containing only single covalent bonds).

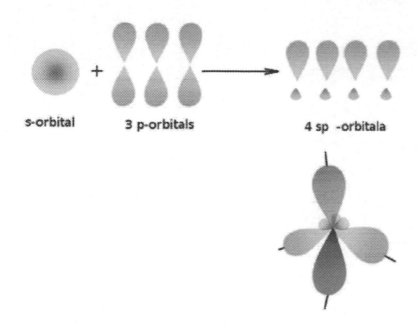

s-orbital   3 p-orbitals   4 sp -orbitala

(b) sp*2* hybridization (involved in organic compounds having carbon atoms linked by double bonds).

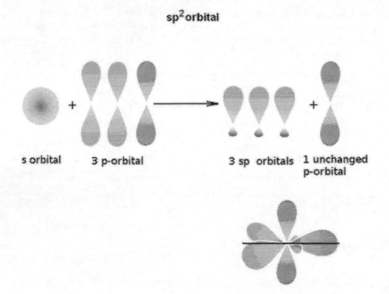

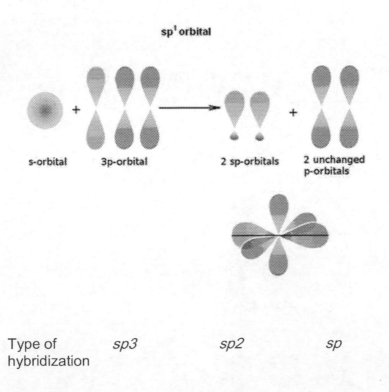

(c) sp hybridization (involved in organic compounds having carbon atoms linked by a triple bonds).

| Type of hybridization | *sp3* | *sp2* | *sp* |

Number of orbitals used	1s and 3p	1s and 2p	1s and 1p
Number of unused p-orbitals	Nil	One	Two
Bond	Four -s	Three –s One -p	Two -s Two -p
Bond angle	109.5°	120°	180°
Geometry	Tetrahedral	Trigonal planar	Linear
% s-character	25 or 1/4	33.33 or 1/3	50 or 1/2

Can you explain the tetrahedral structure of molecules?

The hybridization of sp3 can explain the tetrahedral structure of molecules.

Methane (CH_4) represents the tetrahedral structure molecule as shown below:

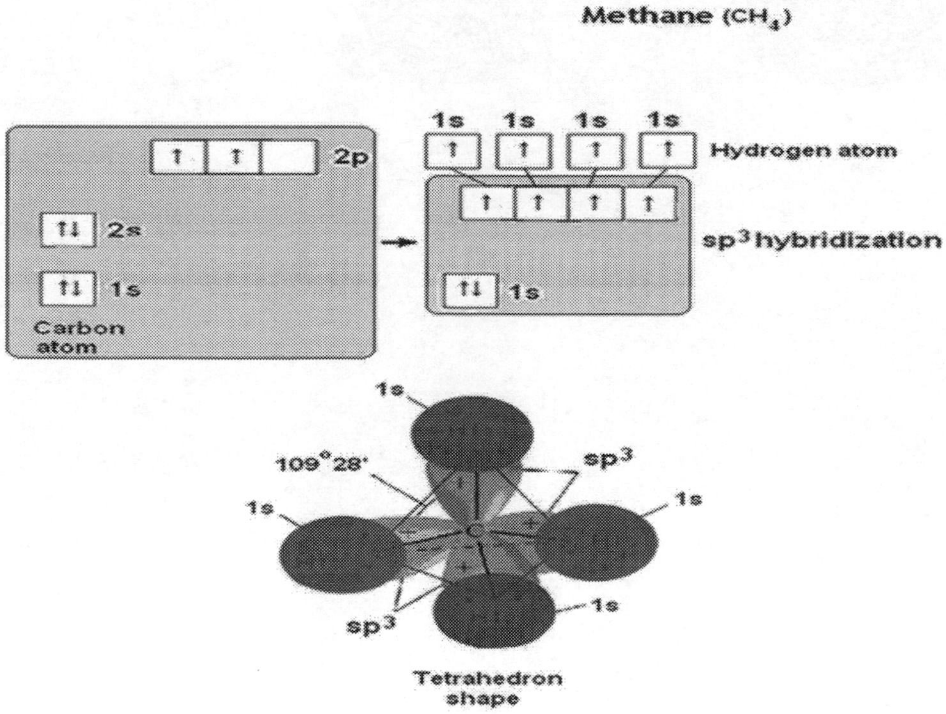

Methane (CH_4)

Can you give another example of sp3?

It is the Dichloroethane (C2H2Cl2).

$C_2H_2Cl_2$

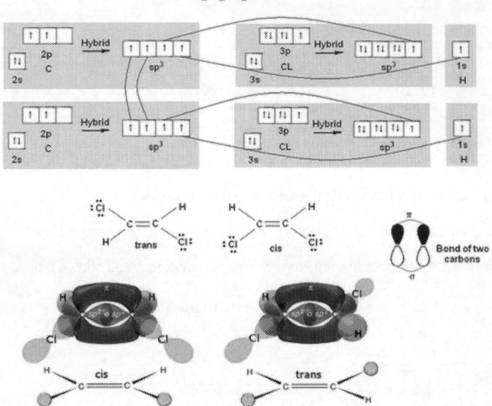

What about the sp2 hybridization?

Formaldehyde (CH2O) is the answer.

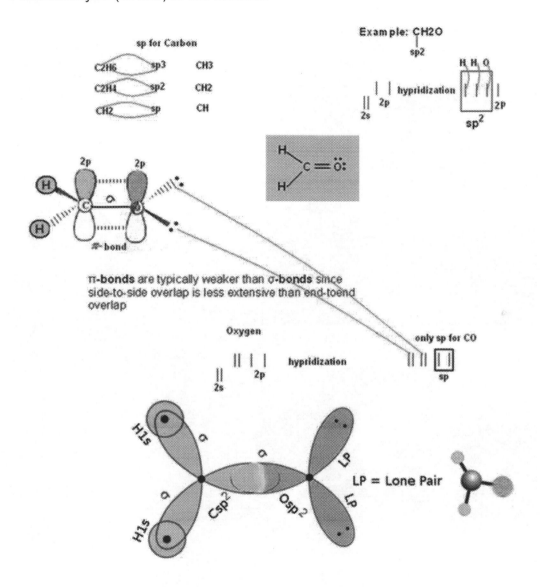

Can you give more examples of sp1?

The sp1 Hybridization can explain the linear structure in molecules. In it, the 2s orbital and one of the 2p orbitals hybridize to form two sp orbitals, each consisting of 50% s and 50% p character. The front lobes face away from each other and form a straight line leaving a 180° angle between the two orbitals.

(a) Magnesium Hydride (MgH_2)

Can you show the hybridization of ethyne (C₂H₂)?

Ethyne (C$_2$H$_2$)

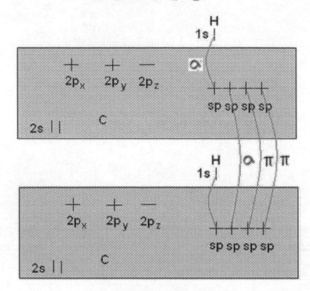

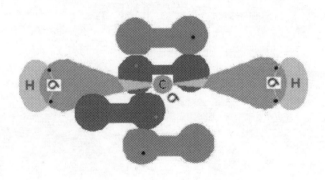

What are the bond angles?

Hybridization	# of Bonds	# of Non-Bonding Pairs	Molecular Shape Geometry	Bond Angles	Examples
sp	2	0	Linear	180°	BeH_2, CO_2
sp^2	3	0	 Trigonal planar	120°	SO_3, BF_3
sp^2	2	1	Angular 	<120°	SO_2, O_3
sp^3	4	0	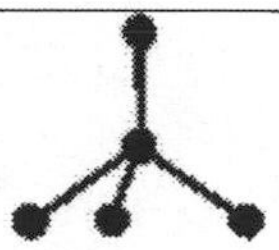 Tetrahedral	109.5°	CH_4, CF_4, SO_4^{2-}
sp^3	3	1	Trigonal pyramidal 	<109.5°	NH_3, PF_3, $AsCl_3$
sp^3	2	2	Bent 	<109.5°	H_2O, H_2S, SF_2
sp^3d	5	0	Trigonal bipyramidal 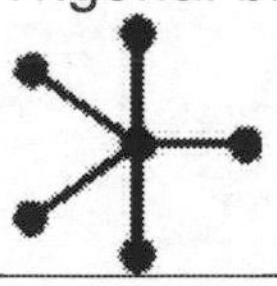	120°, 90°	PF_5, PCl_5, AsF_5
sp^3d	4	1	Sawhorse 	<120°, <90°	SF_4
sp^3d	3	2	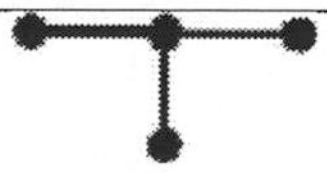 T-shaped	< 90°	ClF_3
sp^3d	2	3	Linear	180°	XeF_2, IF_2, I_3^{1-}
sp^3d^2	6	0	 Octahedral	90°	SF_6, PF_6^{1-}, SiF_6^{2-}
sp^3d^2	5	1	Square pyramidal	< 90°	IF_5, BrF_5

			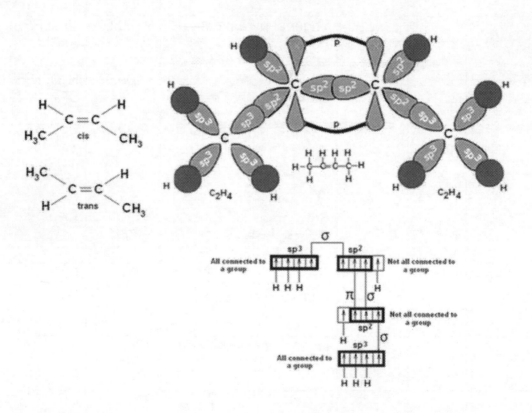			
sp^3d^2	4	2	Square Planar	90 °	XeF_4, IF_4

Can you show some diagrams of mixture of sp^1, sp^2, and sp^3?

(a) 2-butene (C_4H_8)

2-butene (C_4 H_8)

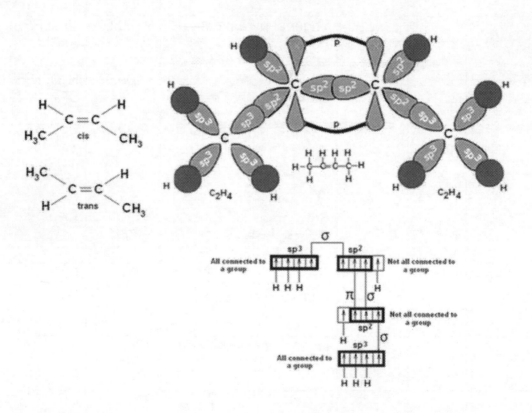

(b) C2H2Cl2

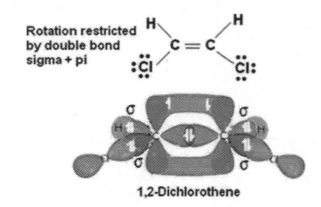

Rotation restricted
by double bond
sigma + pi
H
H
C
C
Cl
Cl
σ
σ
σ
σ
H
H

1,2-Dichlorothene

$C_2H_2Cl_2$

C
H
Cl
C
H
Cl

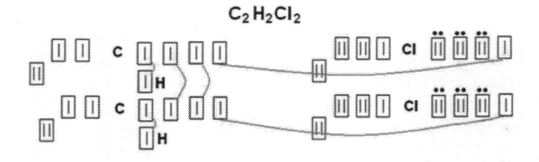

(c) C2H4Cl2

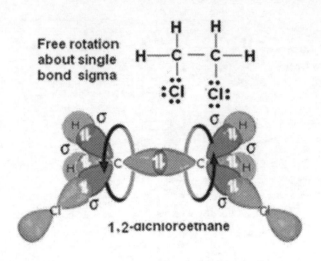

Free rotation about single bond sigma

1,2-dichloroethane

$$C_2H_4Cl_2$$

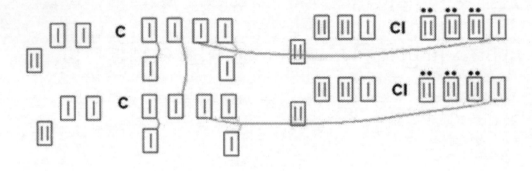

(d) Carbon (sp1, sp2, sp3)

Carbon hybridization
sp3 hybridized
σ
σ
H
C
sp2 hybridized
π
σ
sp hybridized
π
σ
π

(e) Formaldheyde ($CH2O$)

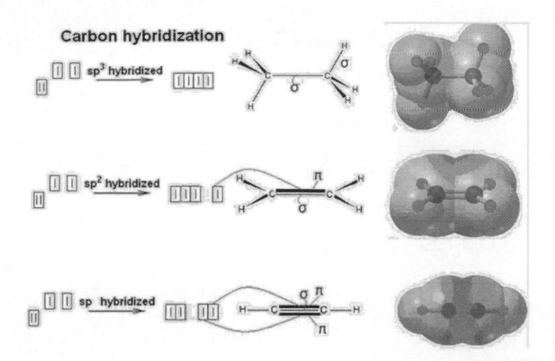

Formaldheyde (CH₂O) orbital diagram

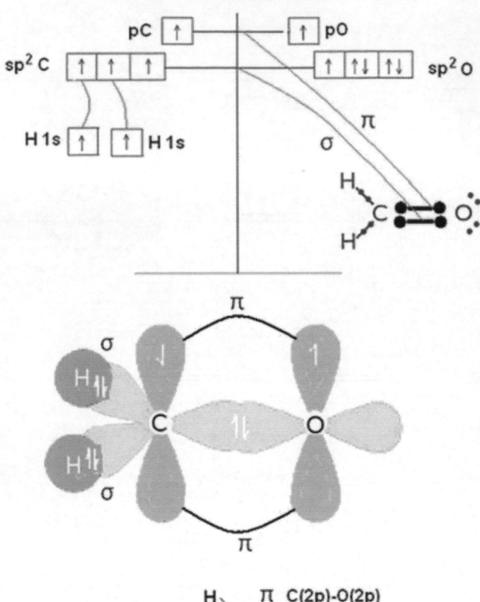

σ: H(1s)-C(sp²)   π C(2p)-O(2p)

σ: C(sp²- O(2p)

(f) Ethane (CH2CH2)

Ethene (CH₂CH₂) orbital diagram

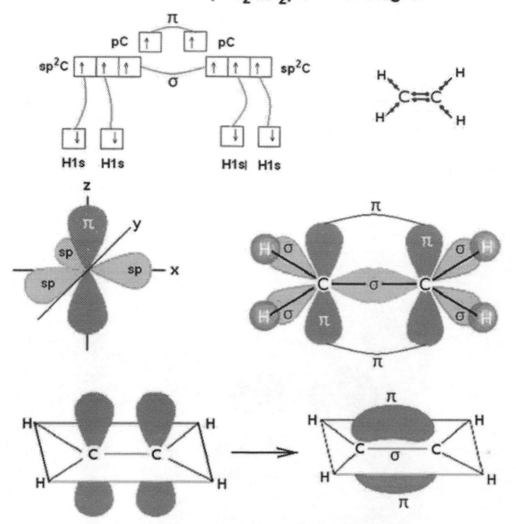

Overlap of p orbitals leading to pi (π) bond

(g) Ethylamine (CH2NH)

Ethylamine (CH₂NH)

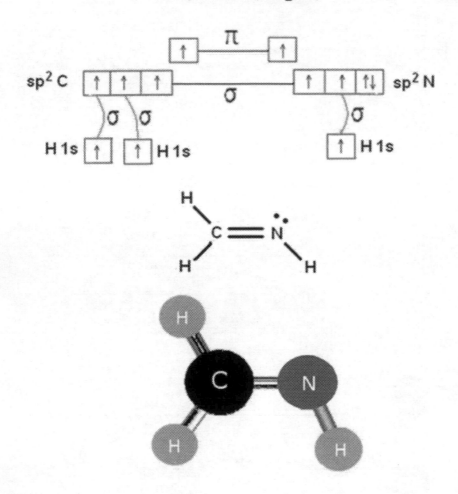

(h) Phosphorus pentachloride PCl5

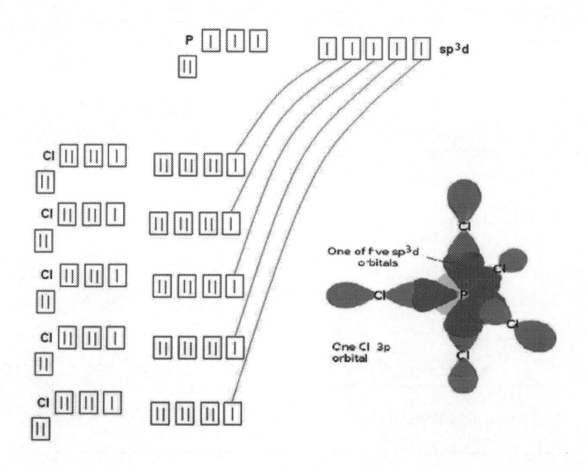

(i) Sulfur and fluorine

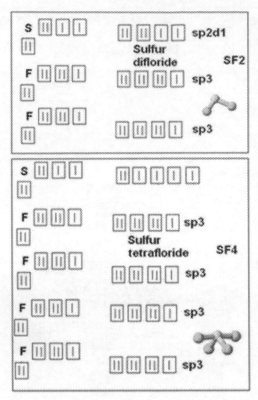

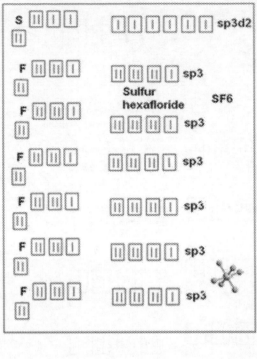

(j) Bromine monofloride (BrF)

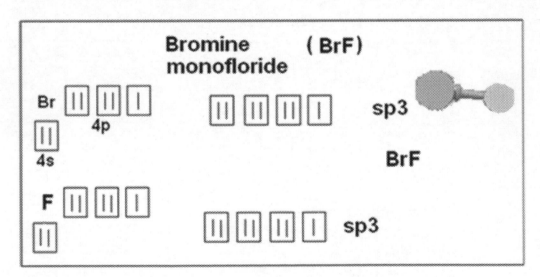

(k) Bromine triflouride (BrF3)

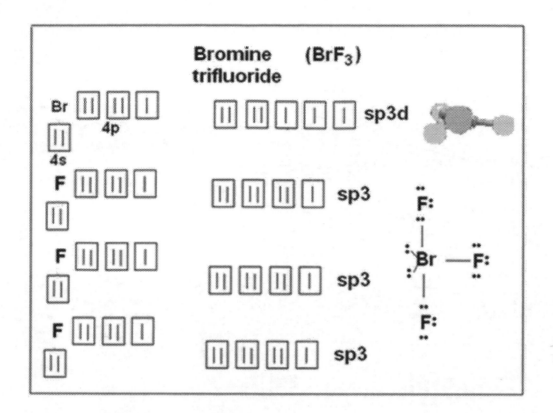

(l) Thionyl tetrafluoride (SOF4)

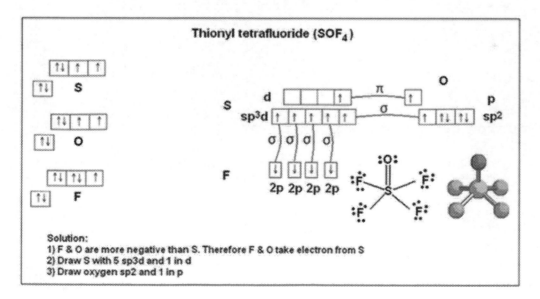

(m) Oxygen (O2 and O)

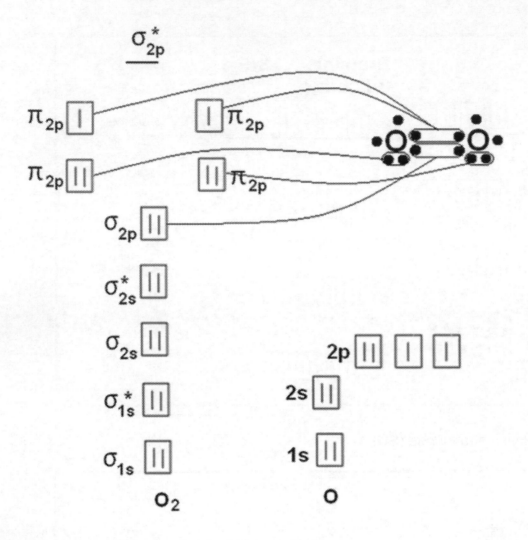

Can you show the level of energy in orbitals?

Before we talk about energy level of molecular bonds, we shall distinguish between the groups B_2, C_2, N_2 and O_2, F_2, Ne_2, as shown in the figure below:

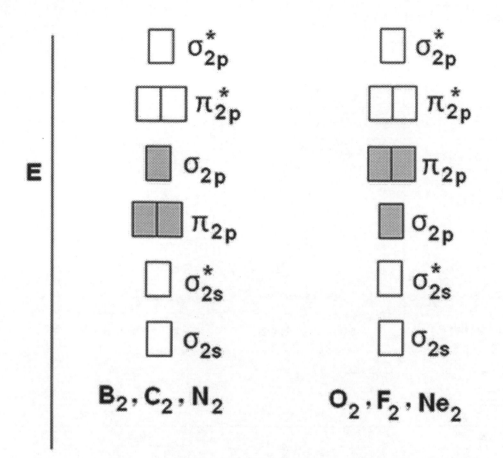

B₂, C₂, N₂   **O₂, F₂, Ne₂**

Note that bonds with* are higher in energy

Examples:

C_2^{2+}, C_2, C_2^{2-}

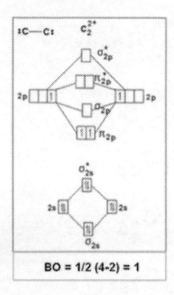

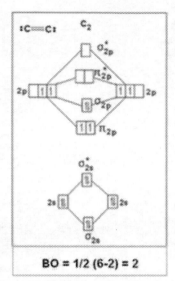

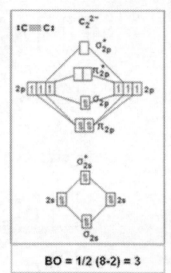

O_2^{2+}, O_2, O_2^{2-}

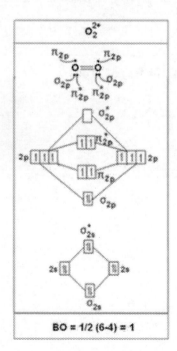

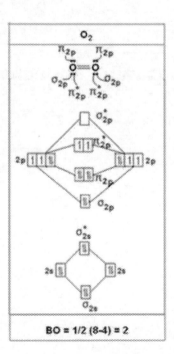

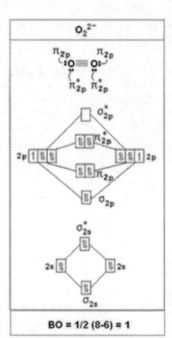

Can you show other molecular orbital energy?

Hydrogen

Hydrogen

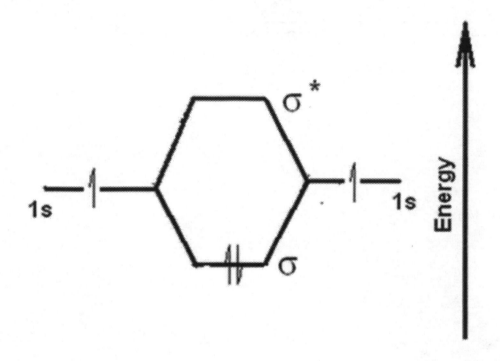

Methane (CH₄)

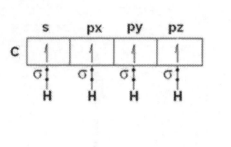

Methane (CH₄)

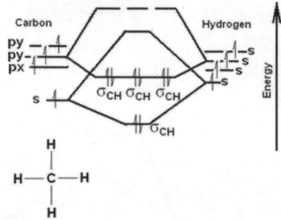

Ethane (C₂H₆)

Ethane (H₃C– C₃H)

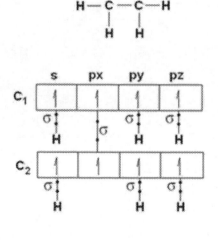

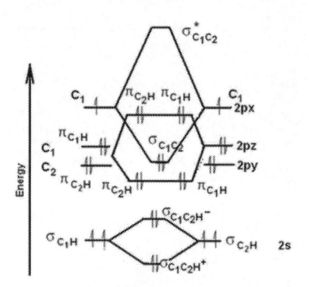

Ethene (C₂H₄)

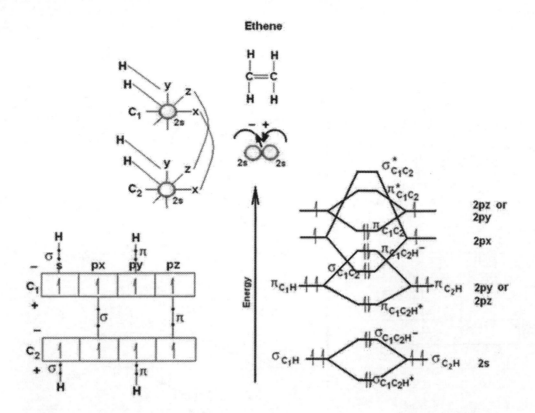

Ethene

One can see that ethane (alkene) is dominated by two "frontier orbitals", that is the Highest Occupied Orbital (HOMO) and the Lowest Occupied Molecular Orbital (LUMO), which are represented by πc1c2 and π*c1c2 respectively. The frontier orbitals don't allow the molecule to rotate around C=C which tends to hold the molecule flat.

Ethyne (C₂H₂)

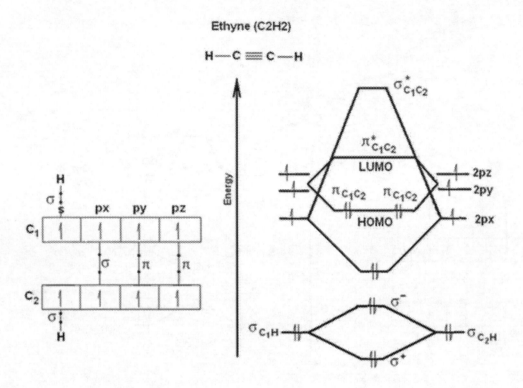

Hydrogen Fluoride (HF)

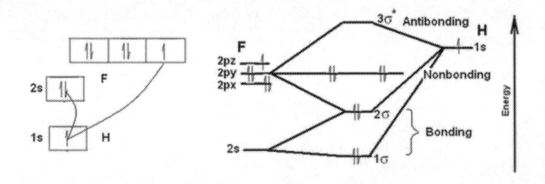

Water (H₂O)

Water (H$_2$O)

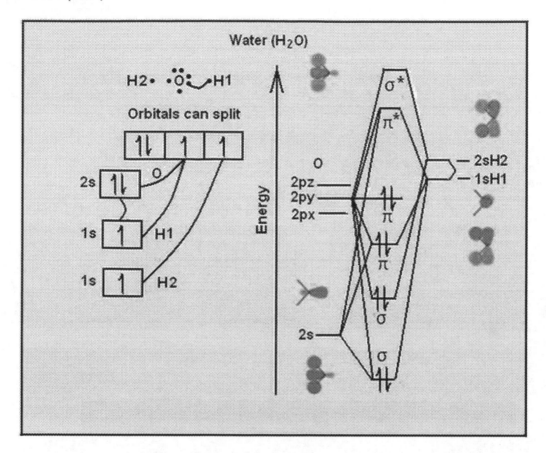

Can you define electronegativity and energy?

Elements with higher electronegativity have lower energy than higher electronegativity as shown in the following examples:

Alkene (C_3H_6)

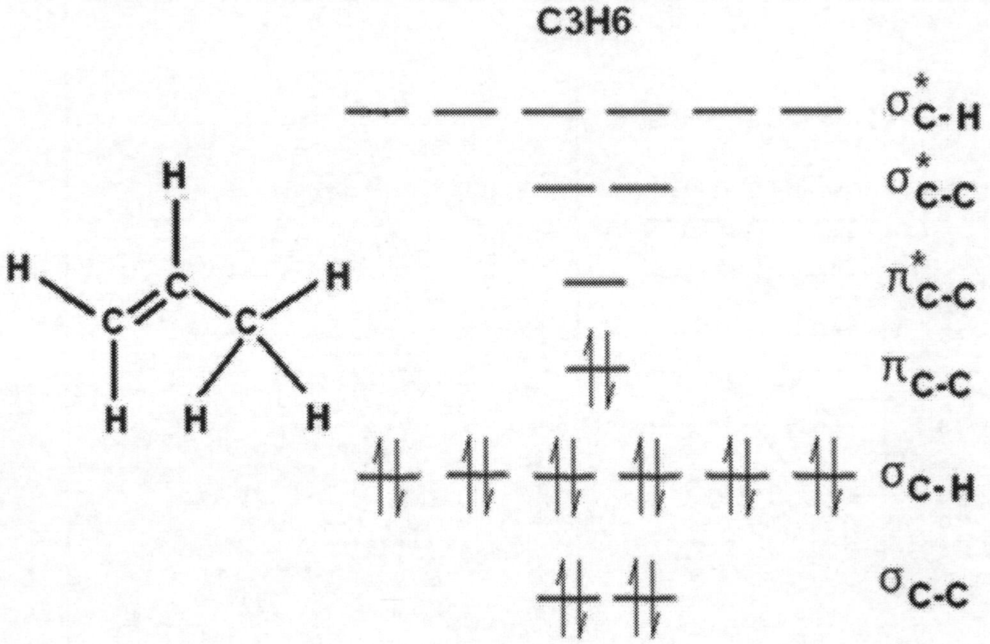

Ethanol (C$_2$H$_6$O)

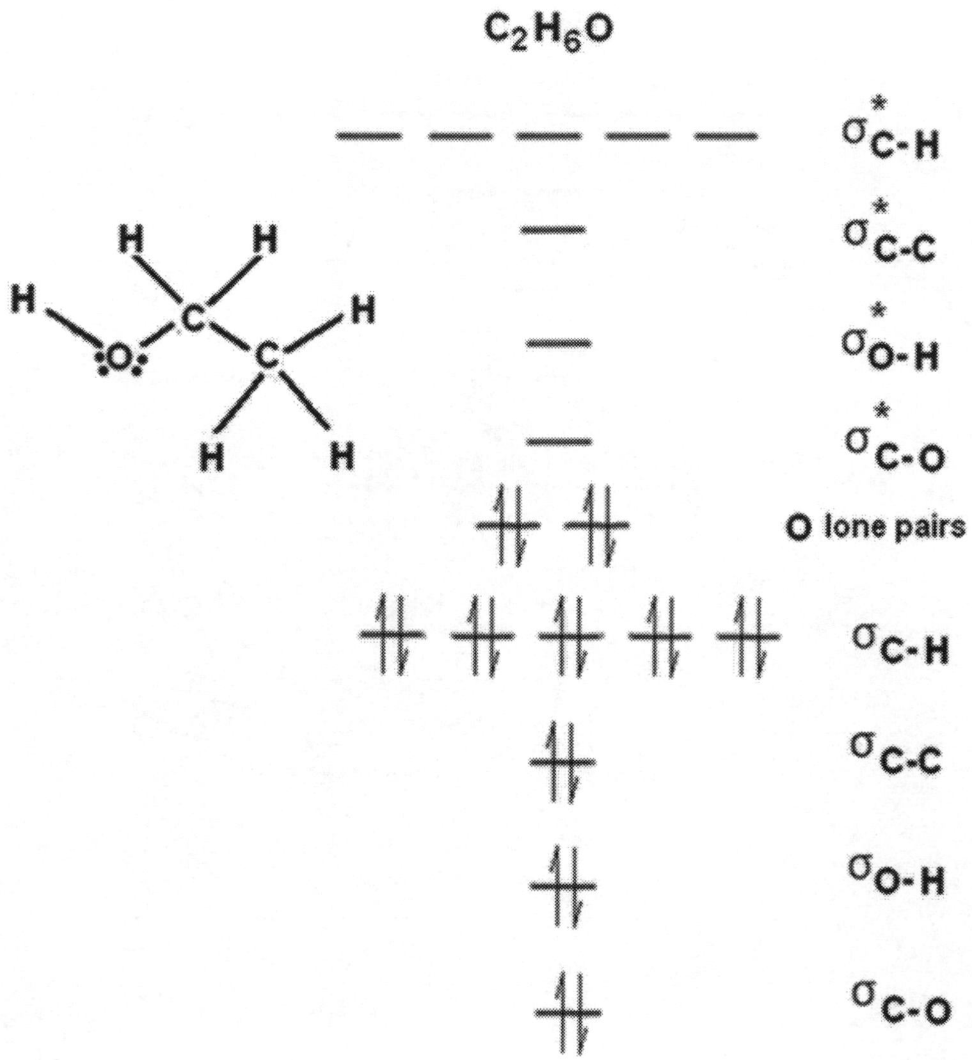

$$C_2H_6O$$

σ^*_{C-H}

σ^*_{C-C}

σ^*_{O-H}

σ^*_{C-O}

O lone pairs

σ_{C-H}

σ_{C-C}

σ_{O-H}

σ_{C-O}

Ethylene Oxide (C$_2$H$_4$O)

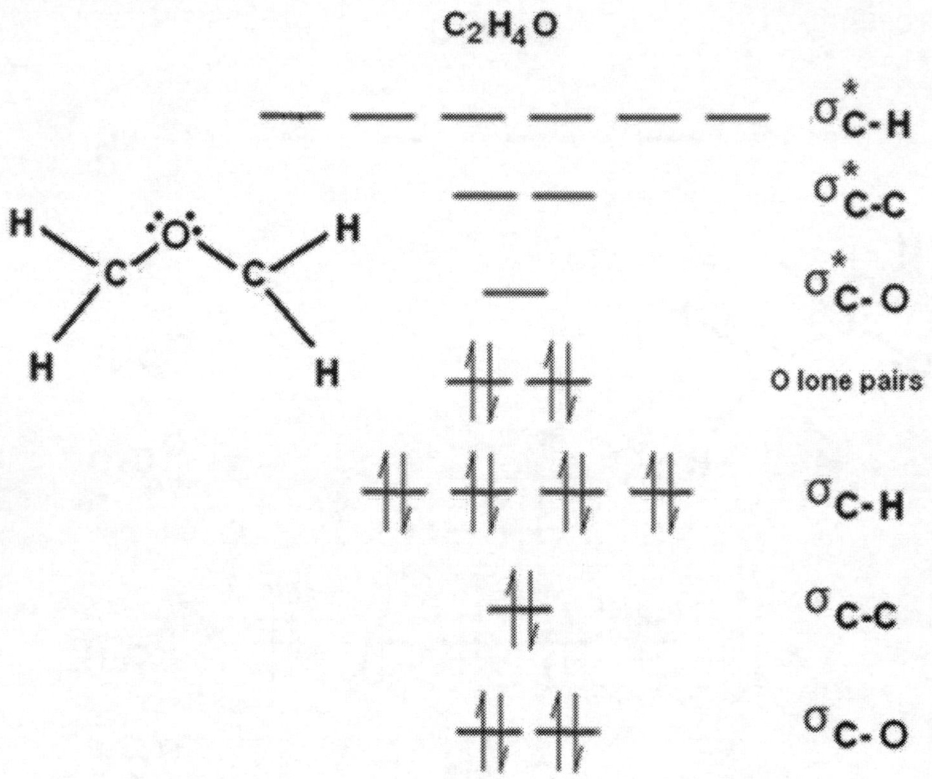

C$_2$H$_4$O

$$\sigma^*_{C-H}$$

$$\sigma^*_{C-C}$$

$$\sigma^*_{C-O}$$

O lone pairs

$$\sigma_{C-H}$$

$$\sigma_{C-C}$$

$$\sigma_{C-O}$$

Cyanogen Bromide (CNBr)

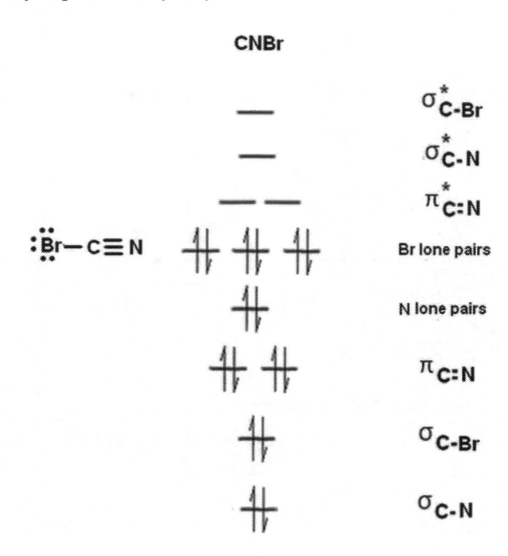

Note: π is higher than σ, where as σ* is higher than π*.

Oxygen Molecule (O2)

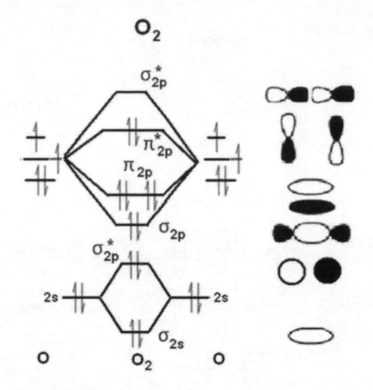

The upper half is not filled ➔ paramagnetic

Bond order = 1/2 (no. of bonding orbitals - no. of antibonding orbitals)

=

1/2 (8-4) = 2

Nitrogen (N2)

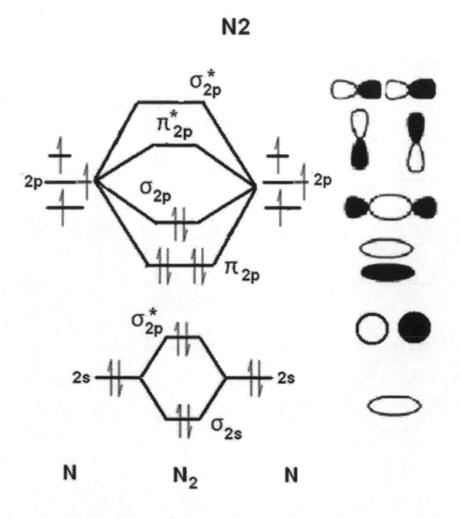

N2

N   N₂   N

No electrons in the upper half. the lower half is filled ➡ Dimagnetic

Bond order = 1/2 (8-2) = 3 (i.e. N≡N)

HOMO = σ_{2p} , Lumo = π^*_{2p}

Note that electronegativity in oxygen is larger than in nitrogen, therefore, larger ΔE between s and p orbitals, thus less orbital mixing. In other words, oxygen has more protons than nitrogen, which pulls the electrons in closer to the nucleus (lower energy). This makes the energy in the s-orbital less than before (before means ground state).

MO Diagram for CO (:C̄≡O:⁺)

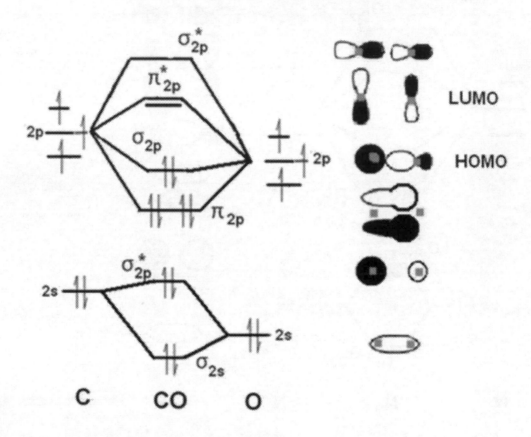

LUMO

HOMO

C   CO   O

MO diagram is similar to that of N₂

CO is diamagnetic
Bond order of 3 C≡O

No electrons in the upper half. the lower half is
filled ⟶ Dimagnetic

Can you define donor and acceptor?

The ordinary covalent bond between two atoms is due to the interaction of two electrons, one from each atom. The donor-acceptor bond is formed by a pair of electrons from one atom (the donor) and a free (unfilled) orbital from another (the acceptor).
A donor is a high energy orbital with one or more electrons. An acceptor is a low energy orbital with one or more vacancies:
A donor is an atom or group of atoms whose highest filled atomic orbital or molecular orbital is higher in energy than that of a reference orbital

An acceptor is an atom or group of atoms whose lowest unfilled atomic or molecular orbital is lower in energy than that of a reference orbital.

Can you define donor and acceptor with respect to Lewis and Bronsted?

Lewis acid: **A** Lone pair acceptor
Lewis base: **:B** Lone pair donor
A and **:B** combine to give **A:B**, the bonding orbital generated is occupied by the electrons supplied by **:B**.

The Lewis definition: an acid is an electron acceptor, and a base is an electron donor.

The Bronsted (or Bronsted-Lowry) definition: an acid is a proton (H+ ion) donor, and a base is a proton acceptor;

Usually, the base has a suitable HOMO (highest occupied molecular orbital) whereas the acid possesses a suitable LUMO (lowest unoccupied molecular orbital). The interaction between the full HOMO of the base and the empty LUMO of the acid gives rise to a bonding and an antibonding orbital as pictured in Figure (52).

Figure (52): HOMO and LOMO

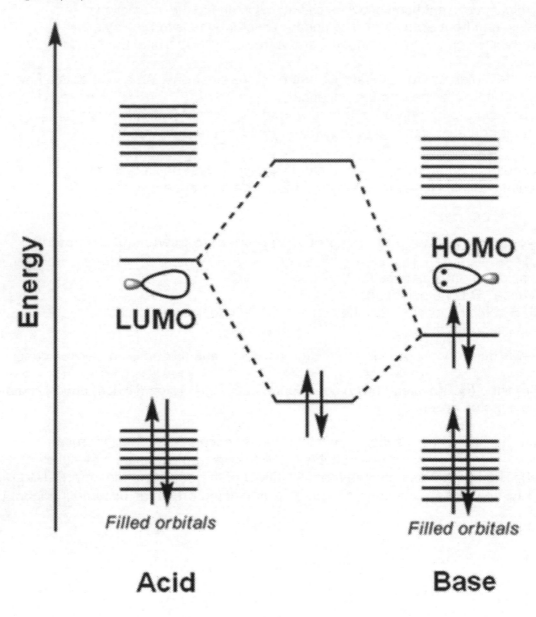

What is the effect of electronegativity on donor and acceptor?

Electronegativity is very important in molecular bonding. Here are two molecular of which their outside orbitals are saturated, where the oxygen (larger electronegativity than Nitrogen's) is a donor, Figure (53).

Figure (53): Effect of electronegativity on bonding

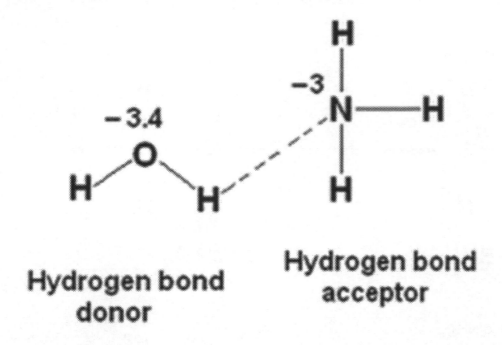

What is the effect of Electrons in the Outermost Shell?

Elements of the 2nd and 3rd periods have fewer than 4 electrons, and therefore the number of covalent bonds, which the given element can form, is equal to the number of electrons in the outermost shell of the given element. For example, lithium (Li), beryllium (Be), and boron (B) can form 1, 2, and 3 covalent bonds respectively. Elements with less than 4 electrons remain unsaturated. For example, Atoms of sodium (Na), magnesium (Mg), and aluminum (Al), after forming the maximal number of covalent bonds, will have 2, 4, and 6 electrons in the outermost shell when they form molecule NaF, MgF2, and AlF3 respectively. Note that metal and nonmetal bonds are ionic bonds.

Elements with more than 4 electrons need 8 electrons to be situated in the outermost shell. Therefore, nitrogen (N), oxygen (O), fluorine (F), and neon (Ne) with 5, 6, 7, and 8 electrons in the outermost shell can form 3, 2, 1, 0 covalent chemical bonds respectively.

Atoms of nitrogen (N), oxygen (O), and fluorine (F), after the formation of 3, 2, and 1 covalent bonds, will contain 8 electrons in the outermost shell, of which the bonding ones are: nitrogen (N) - 6 electrons; oxygen (O) - 4 electrons; fluorine (F) - 2 electrons.

Covalent bonds such as NH3 (ammonium), H2O (water) and HF (hydrogen fluoride) contain 2, 4, and 6 electrons in the outermost shells respectively. Such electrons do not take part in chemical bond formation. They are free electrons, and they are considered to be donors.

Examples of donors and acceptors:

1- H_3B and $N(CH_2)_2$

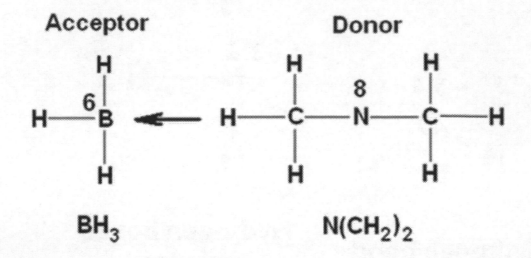

2- H_3B and NH_3

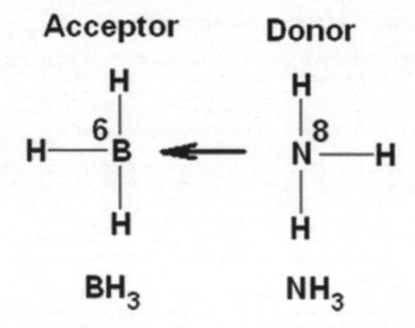

3- $CL3Al$ and $NH3$

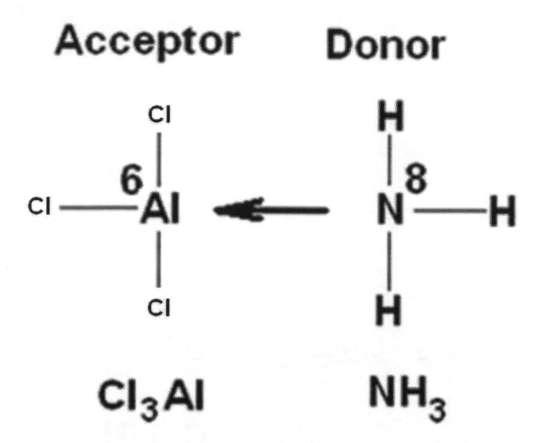

The bonding energy between Al and N in compound Cl3Al←NH3 comprises 165 kJ/mol; while the covalent bonding energy between Al and N is equal to about 400 kJ/mol. This is because their orbits are occupied by more electrons.

4- Cl_2Be and $O(C_2H_5)_2$

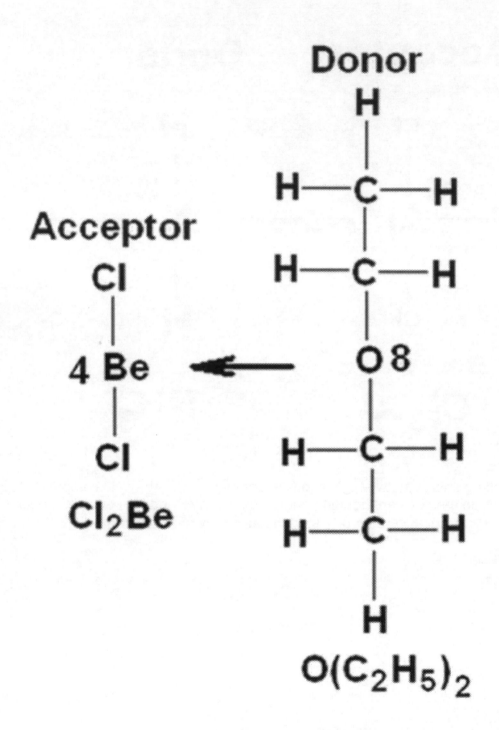

Can you explain π and σ Ligands?

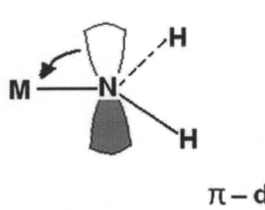 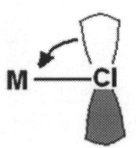

$$\pi - \text{donor}$$

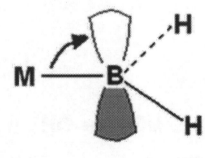 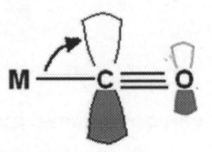

$$\pi - \text{acceptor}$$

Metal + NH₂ ➡ **π – donor**
Metal + Cl ➡ **π – donor**
metal + BH₂ ➡ **π – acceptor**
Metal + C ➡ **π – acceptor**

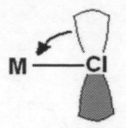

π – donor

C can be replaced by CO, NO⁺, or CN⁻

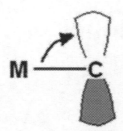

π – acceptor

C can be replaced by Cl⁻, OH⁻, NR₂⁻, or OH₂

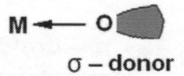

σ – donor

O can be replaced by NH₃, CH₃⁻, or H⁻

O2- and F- stabilize strong Lewis acids (typically elements in high oxidation states) by using full 2Px(y) orbitals as electron-pair donors in addition to the 2pz used in σ-bonding. This is equivalent to the idea that highest oxidation states of an element are manifested in compounds with oxygen.

Glossary

Acid

A substance that produces H+(aq) ions in aqueous solution. Strong acids ionize completely or almost completely in dilute aqueous solution. Weak acids ionize only slightly.

Acid Anhydride

Compound produced by dehydration of a carbonic acid; general formula is R--C--O--C--R

Acidic Salt

A salt containing an ionizable hydrogen atom; does not necessarily produce acidic solutions.

Activation Energy

Amount of energy that must be absorbed by reactants in their ground states to reach the transition state so that a reaction can occur.

Acyl Group

Compound derived from a carbonic acid by replacing the --OH group with a halogen (X), usually --Cl; general formula is O R--C—X

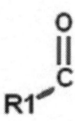

Alcohol

Hydrocarbon derivative containing an --OH group attached to a carbon atom not in an aromatic ring

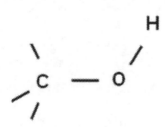

Aldehyde

Compound in which an alkyl or aryl group and a hydrogen atom are attached to a carbonyl group and a hydrogen atom are attached to a carbonyl group; general formula, O-R-C-H

Alkali Metals

Metals of Group IA (Na, K, Rb).

Alkaline Earth Metals

Group IIA metals

Alkenes

Unsaturated hydrocarbons that contain one or more carbon-carbon double bonds.

Alkyl Group

A group of atoms derived from an alkane by the removal of one hydrogen atom.

Alkylbenzene

A compound containing an alkyl group bonded to a benzene ring.

Alkynes

Unsaturated hydrocarbons that contain one or more carbon-carbon triple bonds.

Alpha Particle

A helium nucleus.

Alpha (a) Particle

Helium ion with 2+ charge; an assembly of two protons and two neutrons. Compound that can be considered a derivative of ammonia in which one or more hydrogen are replaced by a alkyl or aryl groups.

Amine

Derivatives of ammonia in which one or more hydrogen atoms have been replaced by organic groups

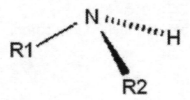

Amine Complexes

Complex species that contain ammonia molecules bonded to metal ions.

Amino Acid

Compound containing both an amino and a carboxylic acid group.The -- NH2 group.

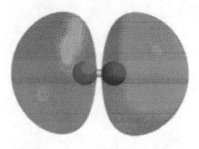

Anion
    A negative ion; an atom or goup of atoms that has gained one or more
    electrons.
Anode
    In a cathode ray tube, the positive electrode.
    Electrode at which oxidation occurs.
Antibonding Orbital
    A molecular orbital higher in energy than any of the atomic orbitals from
    which it is derived; lends instability to a molecule or ion when populated with
    electrons; denoted with a star (*) superscript or symbol.

Aromatic Hydrocarbons
    Benzene and its derivatives.

Aryl Group

Group of atoms remaining after a hydrogen atom is removed from the aromatic system.

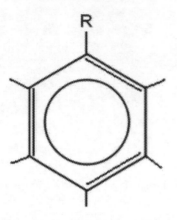

Associated Ions
Short-lived species formed by the collision of dissolved ions of opposite charges.

Atom
The smallest particle of an element

Atomic Mass Unit (amu)
One twelfth of a mass of an atom of the carbon-12 isotope; a unit used for stating atomic and formula weights; also called dalton.

$$A = Z + N$$

A   **atomic mass 14**
Z   **atomic Number 6** C

N   **neutron number 8**

Atomic Number
Integral number of protons in the nucleus; defines the identity of element.

Atomic Orbital
Region or volume in space in which the probability of finding electrons is highest.

Atomic Radius
Radius of an atom.

Atomic Weight
Weighted average of the masses of the constituent isotopes of an element; The relative masses of atoms of different elements.

Aufbau ('building up') Principle
Describes the order in which electrons fill orbitals in atoms.

Band

A series of very closely spaced, nearly continuous molecular orbitals that belong to the crystal as a whole.

Base

A substance that produces OH (aq) ions in aqueous solution. Strong soluable bases are soluble in water and are completely dissociated. Weak bases ionize only slightly.

Basic Anhydride

The oxide of a metal that reacts with water to form a base.

Basic Salt

A salt containing an ionizable OH group.

Beta Particle

Electron emitted from the nucleus when a neuton decays to a proton and an electron.

Binding Energy (nuclear binding energy)

The energy equivalent ($E = mc^2$) of the mass deficiency of an atom.
where: E = is the energy in joules, m is the mass in kilograms, and c is the speed of light in m/s^2

Boiling Point

The temperature at which the vapor pressure of a liquid is equal to the applied pressure; also the condensation point

Boiling Point Elevation

The increase in the boiling point of a solvent caused by the dissolution of a nonvolatile solute.

Bond Energy

The amount of energy necessary to break one mole of bonds of a given kind (in gas phase).

The amount of energy necessary to break one mole of bonds in a substance, dissociating the sustance in the gaseous state into atoms of its elements in the gaseous state.

Bond Order

Half the numbers of electrons in bonding orbitals minus half the number of electrons in antibonding orbitals.

Bonding Orbital

A molecular orbit lower in energy than any of the atomic orbitals from which it is derived; lends stability to a molecule or ion when populated with electron

Bonding Pair

Pair of electrons involved in a covalent bond.

Calorie

The amount of heat required to raise the temperature of one gram of water from 14.5°C to 15.5°C. 1 calorie = 4.184 joules.

Carcinogen

A substance capable of causing or producing cancer in mammals.

Catalyst

A substance that speeds up a chemical reaction without being consumed itself in the reaction.

A substance that alters (usually increases) the rate at which a reaction occurs.

Cathode

Electrode at which reduction occurs

In a cathode ray tube, the negative electrode.

Cathode Ray Tube

Closed glass tube containing a gas under low pressure, with electrodes near the ends and a luminescent screen at the end near the positive electrode; produces cathode rays when high voltage is applied.

Cation

A positive ion; an atom or group of atoms that has lost one or more electrons.

Cell Potential

Potential difference, Ecell, between oxidation and reduction half-cells under nonstandard conditions.

Central Atom

An atom in a molecule or polyatomic ion that is bonded to more than one other atom.

Chain Reaction

A reaction that, once initiated, sustains itself and expands.

This is a reaction in which reactive species, such as radicals, are produced in more than one step. These reactive species, radicals, propagate the chain reaction.

Chemical Bonds

The attractive forces that hold atoms together in elements or compounds.

Chemical Change

A change in which one or more new substances are formed.

Chemical Equation

Description of a chemical reaction by placing the formulas of the reactants on the left and the formulas of products on the right of an arrow.

Chemical Equilibrium

A state of dynamic balance in which the rates of forward and reverse reactions are equal; there is no net change in concentrations of reactants or products while a system is at equilibrium.

Chemical Periodicity

The variations in properties of elements with their position in the periodic table

Cis-

The prefix used to indicate that groups are located on the same side of a bond about which rotation is restricted.

H3C   CH3
   \   /
    C == C
   /   \
  H   H
      cis

  H   CH3
   \   /
    C == C
   /   \
 H3C   H
      trans

Cis-Trans Isomerism
    A type of geometrical isomerism related to the angles between like ligands.
Colloid
    A heterogeneous mixture in which solute-like particles do not settle out.
Combination Reaction
    Reaction in which two substances (elements or compounds) combine to form one compound.
    Reaction of a substance with oxygen in a highly exothermic reaction, usually with a visible flame.
Complex Ions
    Ions resulting from the formation of coordinate covalent bonds between simple ions and other ions or molecules.
Compound
    A substance of two or more elements in fixed proportions. Compounds can be decomposed into their constituent elements.
    For more Information.
Coordinate Covalent Bond
    Covalent bond in which both shared electrons are furnished by the same species; bond between a Lewis acid and Lewis base.
Coordinate Covalent Bond
    A covalent bond in which both shared electrons are donated by the same atom; a bond between a Lewis base and a Lewis acid.
Coordination Compound or Complex
    A compound containing coordinate covalent bonds.
Coordination Isomers
    Isomers involving exchanges of ligands between complex cation and complex anion of the same compound.
Coordination Number

In describing crystals, the number of nearest neighbours of an atom or ion. The number of donor atoms coordinated to a metal.

Covalent Bond
Chemical bond formed by the sharing of one or more electron pairs between two atoms.

Covalent Compounds
Compounds containing predominantly covalent bonds.

Critical Mass
The minimum mass of a particular fissionable nuclide in a given volume required to sustain a nuclear chain reaction.

Debye
The unit used to express dipole moments.

Density
Mass per unit Volume: D=MV

Deposition
The direct solidification of a vapor by cooling; the reverse of sublimation.

Deuterium
An isotope of hydrogen whose atoms are twice as massive as ordinary hydrogen;deuterion atoms contain both a proton and a neutron in the nucleus.

Hydrogen   Deuterium   Tritium

Dimer
Molecule formed by combination of two smaller (identical) molecules.

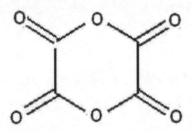

Dipole
Refers to the separation of charge between two covalently bonded atoms

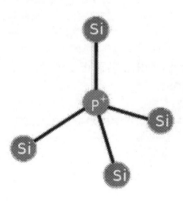

Dipole-dipole Interactions

Attractive interactions between polar molecules, that is, between molecules with permanent dipoles.

$$\sigma^+ \quad \sigma^- \qquad \sigma^+ \qquad \sigma^-$$
$$H\text{———}Cl\text{------}H\text{———}Cl$$

Dipole Moment

The product of the distance separating opposite charges of equal magnitude of the charge; a measure of the polarity of a bond or molecule; a measured dipole moment refers to the dipole moment of an entire molecule.

Donor Atom

A ligand atom whose electrons are shared with a Lewis acid.

D-Orbitals

Beginning in the third energy level, aset of five degenerate orbitals per energy level, higher in energy than s and p orbitals of the same energy level.

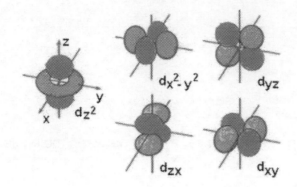

Double Bond
Covalent bond resulting from the sharing of four electrons (two pairs) between two atoms.

D -Transition elements (metals)
B Group elements except IIB in the periodic table; sometimes called simply transition elements EX. Fe, Ni, Cu, Ti .
For further information.

Effective Nuclear Charge
The nuclear charge experienced by the outermost electrons of an atom; the actual nuclear charge minus the effects of shielding due to inner-shell electrons.
Example: Set of $dx_2\text{-}y_2$ and dz_2 orbitals; those d orbitals within a set with lobes directed along the x-, y-, and z-axes.

Electrical Conductivity
Ability to conduct electricity.

Electrochemistry
Study of chemical changes produced by electrical current and the production of electricity by chemical reactions.

Electrodes
Surfaces upon which oxidation and reduction half-reactions; occur in electrochemical cells.

Electrode Potentials
Potentials, E, of half-reactions as reductions versus the standard hydrogen electrode.

Electrolysis
Process that occurs in electrolytic cells.

Electrolyte
A substance whose aqueous solutions conduct electricity.

Electrolytic Cells
Electrochemical cells in which electrical energy causes nospontaneous redox reactions to occur.
An electrochemical cell in which chemical reactions are forced to occur by the application of an outside source of electrical energy.

Electrolytic Conduction
Conduction of electrical current by ions through a solution or pure liquid.

Electromagnetic Radiation

Energy that is propagated by means of electric and magnetic fields that oscillate in directions perpendicular to the direction of travel of the energy.

Electromotive Series

The relative order of tendencies for elements and their simple ions to act as oxidizing or reducing agents; also called the activity series.

Electron

A subatomic particle having a mass of 0.00054858 amu and a charge of 1-.

Electron Affinity

The amount of energy absorbed in the process in which an electron is added to a neutral isolated gaseous atom to form a gaseous ion with a 1- charge; has a negative value if energy is released.

Electron Configuration

Specific distribution of electrons in atomic orbitals of atoms or ions.

Electron Deficient Compounds

Compounds that contain at least one atom (other than H) that shares fewer than eight electrons

Electronic Transition

The transfer of an electron from one energy level to another.

Electronegativity

A measure of the relative tendency of an atom to attract electrons to itself when chemically combined with another atom.

Electronic Geometry

The geometric arrangement of orbitals containing the shared and unshared electron pairs surrounding the central atom of a molecule or polyatomic ion.

Element

A substance that cannot be decomposed into simpler substances by chemical means.

Energy

The capacity to do work or transfer heat.

Enthalpy

The heat content of a specific amount of substance; defined as E= PV. Enthalpy (H) is the sum of the internal energy (U) and the product of pressure and volume (PV) given by the equation: $H = U + PV$

Entropy

A thermodynamic state or property that measures the degree of disorder or randomness of a system. It is the measure of a system's thermal energy per unit temperature: Entropy (Q) = H/T

Enzyme

A protein that acts as a catalyst in biological systems.

Equation of State

An equation that describes the behavior of matter in a given state; the van der Waals equation describes the behavior of the gaseous state.

Equilibrium or Chemical Equilibrium

A state of dynamic balance in which the rates of forward and reverse reactions are equal; the state of a system when neither forward or reverse reaction is thermodynamically favored.

Equilibrium Constant

A quantity that characterizes the position of equilibrium for a reversible reaction; its magnitude is equal to the mass action expression at equilibrium. K varies with temperature.

Equivalence Point

The point at which chemically equivalent amounts of reactants have reacted.

Ester

A Compound of the general formula R-C-O-R1 where R and R1 may be the same or different, and may be either aliphatic or aromatic.

$$R-\overset{\overset{\displaystyle O}{\|}}{C}-O-R'$$

Ether

Compound in which an oxygen atom is bonded to two alkyl or two aryl groups, or one alkyl and one aryl group.

$$-\overset{|}{\underset{|}{C}}-O-\overset{|}{\underset{|}{C}}-$$

Excited State

Any state other than the ground state of an atom or molecule.

Exothermic

Describes processes that release heat energy.

Exothermicity

The release of heat by a system as a process occurs.

Fast Neutron

A neutron ejected at high kinetic energy in a nuclear reaction.

Fat

Solid triester of glycerol and (mostly) saturated fatty acids.

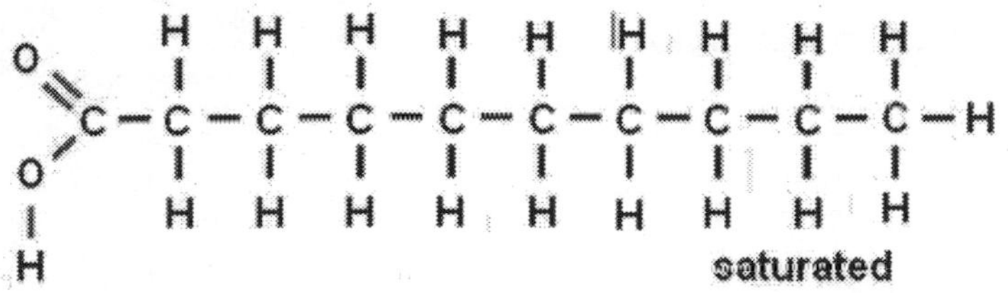

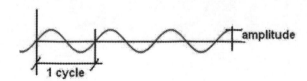

saturated

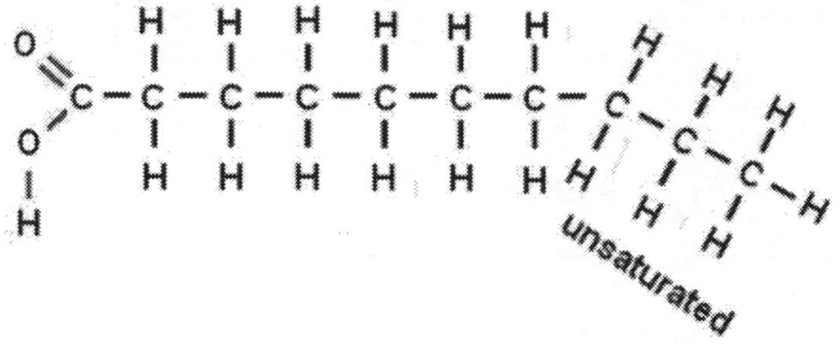

unsaturated

Fatty Acids
      An aliphatic acid; many can obtained from animal fats.

Free Energy Change
      The indicator of spontaneity of a process at constnt T and P. If delta-G is negative, the process is spontaneous.

Free Radical
      A highly reactive chemical species carrying no charge and having a single unpaired electron in an orbital.

Frequency
      The number of repeating corresponding points on a wave that pass a given observation point per unit time.

amplitude

1 cycle

Functional Group
      A group of atoms that represents a potential reaction site in an organic compound.

Functional Groups

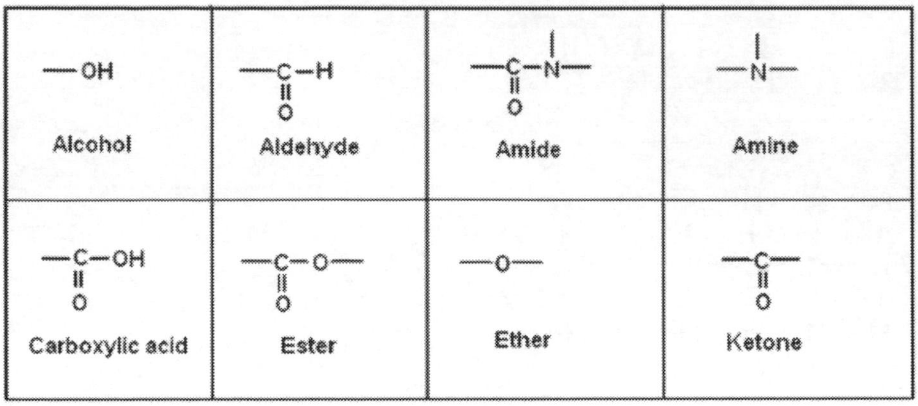

—OH **Alcohol**	—C—H (‖O) **Aldehyde**	—C—N— (‖O) **Amide**	—N— **Amine**
—C—OH (‖O) **Carboxylic acid**	—C—O— (‖O) **Ester**	—O— **Ether**	—C— (‖O) **Ketone**

Gamma Ray

High energy electromagnetic radiation.
A highly penetrating type of nuclear radiation similar to x-ray radiation,
except that it comes from within the nucleus of an atom and has a higher
energy. Energywise, very similar to cosmic ray except that cosmic rays
originate from outer space.

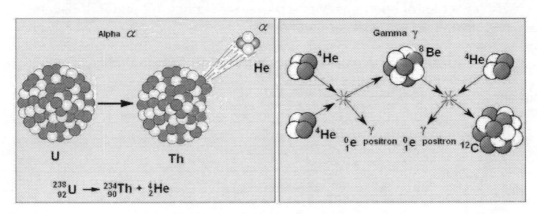

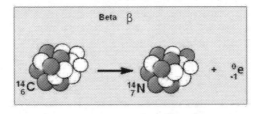

Ground State

The lowest energy state or most stable state of an atom, molecule or ion.

Group

A vertical column in the periodic table; also called a family.

Haber Process

A process for the catalyzed industrial production of ammonia from N_2 and
H_2 at high temperature and pressure.

Half-Life

> The time required for half of a reactant to be converted into product(s).

> The time required for half of a given sample to undergo radioactive decay.

Half-Reaction

> Either the oxidation part or the reduction part of a redox reaction.

Halogens

> Group VIIA elements: F, Cl, Br, I

Hund's Rule

> All orbitals of a given sublevel must be occupied by single electrons before pairing begins.

Hybridization

> Mixing a set of atomic orbitals to form a new set of atomic orbitals with the same total electron capacity and with properties and energies intermediate between those of the original unhybridized orbitals.

Hydration

> Reaction of a substance with water.

Hydration Energy

> The energy change accompanying the hydration of a mole of gase and ions.

Hydride

> A binary compound of hydrogen.

Hydrocarbons

> Compounds that contain only carbon and hydrogen.

Hydrogen Bond

> A fairly strong dipole-dipole interaction (but still considerably weaker than the covalent or ionic bonds) between molecules containing hydrogen directly bonded to a small, highly electronegative atom, such as N, O, or F.

Hydrogenation

> The reaction in which hydrogen adds across a double or triple bond.

Hydrogen-Oxygen Fuel Cell

> Fuel cell in which hydrogen is the fuel (reducing agent) and oxygen is the oxidizing agent.

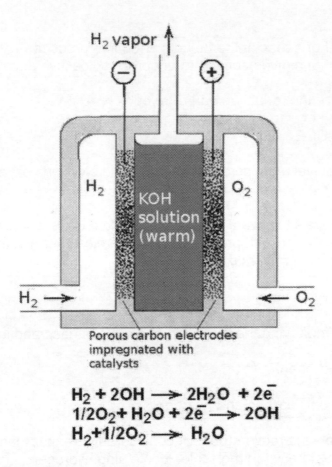

$$H_2 + 2OH \longrightarrow 2H_2O + 2e^-$$
$$1/2O_2 + H_2O + 2e^- \longrightarrow 2OH$$
$$H_2 + 1/2O_2 \longrightarrow H_2O$$

Hydrolysis
    The reaction of a substance with water or its ions.

Inner Orbital Complex
    Valence bond designation for a complex in which the metal ion utilizes d
    orbitals for one shell inside the outermost occupied shell in its hybridization.

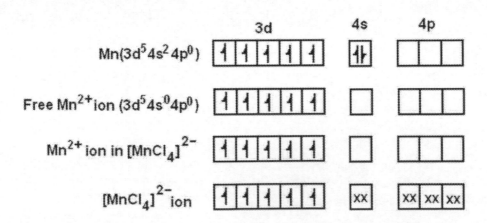

Isomers

Different substances that have the same formula.

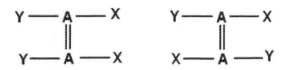

Ionization Isomers
>   Isomers that result from the interchange of ions inside and outside the
>   coordination sphere.

Ionization Constant
>   Equilibrium constant for the ionization of a weak electrolyte.

Ionization
>   The breaking up of a compound into separate ions.

Ideal Gas
>   A hypothetical gas that obeys exactly all postulates of the kinetic-molecular
>   theory.

Ideal Gas Law
>   The product of pressure and the volume of an ideal gas is directly
>   proportional to the number of moles of the gas and the absolute
>   temperature.

> $$PV = n\,RT$$

>   Where P is the pressure of the gas, V is the volume of the gas, n is the
>   amount of substance of gas (also known as number of moles), T is the
>   temperature of the gas and R is the ideal, or universal, gas constant.

Ionization (another definition)
>   In aqueous solution, the process in which a molecular compound reacts
>   with water and forms ions.

Ionic Bonding
>   Chemical bonding resulting from the transfer of one or more electrons from
>   one atom or a group of atoms to another. Ionic bonds are formed between a
>   cation, which is usually a metal, and an anion, which is usually a nonmetal.
>   Pure ionic bonding cannot exist: all ionic compounds have some degree of
>   covalent bonding. Thus, an ionic bond is considered a bond where the ionic
>   character is greater than the covalent character. The larger the difference in
>   electronegativity between the two atoms involved in the bond, the more
>   ionic (polar) the bond is.

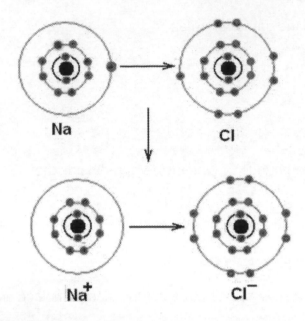

Ionic Compunds
    Compounds containing predominantly ionic bonding.
Ionic Geometry
    The arrangement of atoms (not lone pairs of electrons) about the central atom of a polyatomic ion.
Isoelectric
    Having the same electronic configurations
Ionization Energy
    The minimum amount of energy required to remove the most loosely held electron of an isolated gaseous atom or ion.
Isotopes
    Two or more forms of atoms of the same element with different masses; atoms containing the same number of protons but different numbers of neutrons.

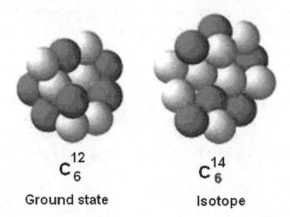

C^{12}_6 C^{14}_6
Ground state   Isotope

Ion
    An atom or a group of atoms that carries an electric charge.

Joule

A unit of energy in the SI system. One joule is 1 kg. m2/s2 which is also 0.2390 calorie.

Ketone

Compound in which a carbonyl group is bound to two alkyl or two aryl groups, or to one alkyl and one aryl group.

Kinetic Energy

Energy that matter processes by virtue of its motion.

Lewis Acid

Any species that can accept a share in an electron pair.

Lewis Base

Any species that can make available a share in an electron pair.

Lewis Dot Formula (Electron Dot Formula)

Representation of a molecule, ion or formula unit by showing atomic symbols and only outer shell electrons

Ligand

A Lewis base in a coordination compound.

Lone Pair

Pair of electrons residing on one atom and not shared by other atoms; unshared pair.

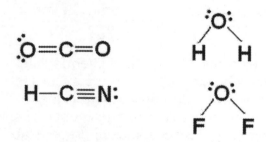

Magnetic Quantum Number (mc)

Quantum mechanical solution to a wave equation that designates the particular orbital within a given set (s, p, d, f) in which a electron resides.

Mass

A measure of the amount of matter in an object. Mass is usually measured in grams or kilograms.

Mass Number

The sum of the numbers of protons and neutrons in an atom; an integer.

Matter

Anything that has mass and occupies space.

Metal

An element below and to the left of the stepwise division (metalloids) in the upper right corner of the periodic table; about 80% of the known elements are metals.

Metallic Bonding
>   Bonding within metals due to the electrical attraction of positively charges
>   metal ions for mobile electrons that belong to the crystal as a whole.

Molecular Formula
>   Formula that indicates the actual number of atoms present in a molecule of
>   a molecular substance.

Molecular Orbital
>   An orbit resulting from overlap and mixing of atomic orbitals on different
>   atoms. An MO belongs to the molecule as a whole.

Molecular Orbital Theory
>   A theory of chemical bonding based upon the postulated existence of
>   molecular orbitals.

Molecular Weight
>   The mass of one molecule of a nonionic substance in atomic mass units.

Molecule
>   The smallest particle of an element or compound capable of a stable,
>   independent existence.

Neutron
>   A neutral subatomic particle having a mass of 1.0087 amu.

Nuclear Binding Energy
>   Energy equivalent of the mass deficiency; energy released in the formation
>   of an atom from the subatomic particles.

Nuclear Fission
>   The process in which a heavy nucleus splits into nuclei of intermediate
>   masses and one or more protons are emitted. Because three super-fast
>   neutrons are produced to every single split U-235 atom, there is the
>   potential for the reaction rate to increase threefold with each bunch of split
>   U-235 atoms, thus more energy (heat) is produced. Nuclear power stations
>   use nuclear fission reaction to super-heat steam in order to drive turbines,
>   as in a conventional power station. That would look like this:

Nuclear Reaction

Involves a change in the composition of a nucleus and can evolve or absorb an extraordinarily large amount of energy

Nuclear Reactor

A system in which controlled nuclear fisson reactions generate heat energy on a large scale, which is subsequently converted into electrical energy.

Nucleons

Particles comprising the nucleus; protons and neutrons.

Nucleus

The very small, very dense, positively charged center of an atom containing protons and neutrons, as well as other subatomic particles.

Nuclides

Refers to different atomic forms of all elements in contrast to ?isotopes?, which refer only to different atomic forms of a single element.

Octet Rule

Many representative elements attain at least a share of eight electrons in their valence shells when they form molecular or ionic compounds; there are some limitations.

Oxidation

An algebraic increase in the oxidation number; may correspond to a loss of electrons.

Oxidation Numbers

Arbitrary numbers that can be used as mechanical aids in writing formulas and balancing equations; for single- atom ions they correspond to the charge on the ion; more electronegative atoms are assigned negative oxidation numbers (also called Oxidation states).

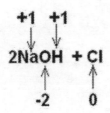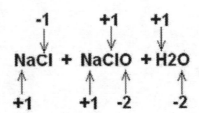

Oxidation-reduction Reactions

Reactions in which oxidation and reduction occur; also called redox reactions.

Oxide

A binary compound of oxygen.

Oxidizing Agent

The substance that oxidizes another substance and is reduced.

Pairing

A favourable interaction of two electrons with opposite m , values in the same orbital.

Pairing Energy

Energy required to pair two electrons in the same orbital.

Particulate Matter

Fine divided solid particles suspended in polluted air.

Pauli Exclusion Principle

No two electrons in the same atom may have identical sets of four quantum numbers.

An orbital can hold 0, 1, or 2 electrons only, and if there are two electrons in the orbital, they must have opposite (paired) spins.

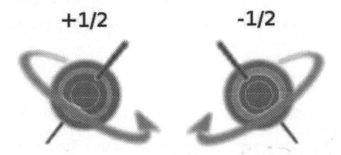

+1/2   -1/2

When we draw electrons, we use up and down arrows. So, if an electron is paired up in a box, one arrow is up and the second must be down.

(Therefore, no two electrons in the same atom can have the same set of four Quantum Numbers (principle quantum number (n=1,2,3,..), orbital quantum number (ℓ =0,1,2...n-1), magnetic quantum number (m_ℓ = - ℓ, - ℓ+1, ...0, ...ℓ -1, ℓ or 2ℓ+1) and spin quantum number (m_s = +1/2, -1/2)).

Percentage Ionization
    The percentage of the weak electrolyte that ionizes in a solution of given concentration.
Period
    The elements in a horizontal row of the periodic table.
Periodicity
    Regular periodic variations of properties of elements with atomic number (and position in the periodic table).
Periodic Law
    The properties of the elements are periodic functions of their atomic numbers.
Periodic Table
    An arrangement of elements in order of increasing atomic numbers that also emphasizes periodicity.
Peroxide
    A compound containing oxygen in the -1 oxidation state. Metal peroxides contain the peroxide ion, O_{22}^-

$$[O—O]^{-2}$$

pH
    Negative logarithm of the concentration (mol/L) of the $H_3O^+[H^+]$ ion; scale is commonly used over a range 0 to 14.
Phenol
    Hydrocarbon derivative containing an [OH] group bound to an aromatic ring.

Photon
A packet of light or electromagnetic radiation; also called quantum of light

Polar Bond
Covalent bond in which there is an unsymmetrical distribution of electron density.

Polarization
The buildup of a product of oxidation or a reduction of an electrode, preventing further reaction.

Positron
A Nuclear particle with the mass of an electron but opposite charge.

Potential Energy
Energy that matter possesses by virtue of its position, condition or composition.

Proton
A subatomic particle having a mass of 1.0073 amu and a charge of +1, found in thew nuclei of atoms.

Quantum Mechanics
Mathematical method of treating particles on the basis of quantum theory, which assumes that energy (of small particles) is not infinitely divisible.

Quantum Numbers
Numbers that describe the energies of electrons in atoms; derived from quantum mechanical treatment.

Radiation
High energy particles or rays emitted during the nuclear decay processes.

Radical
An atom or group of atoms that contains one or more unpaired electrons (usually very reactive species).

$$8\text{-}4\text{-}3 = 1 \qquad 8\text{-}6\text{-}1 = 1$$
$$\overset{-}{C} \equiv \overset{+}{N} - \overset{-}{O}$$
$$8\text{-}5\text{-}4 = -1$$

Radioactive Dating
Method of dating ancient objects by determining the ratio of amounts of mother and daughter nuclides present in an object and relating the ratio to the object? Its age will be via half-life calculations.

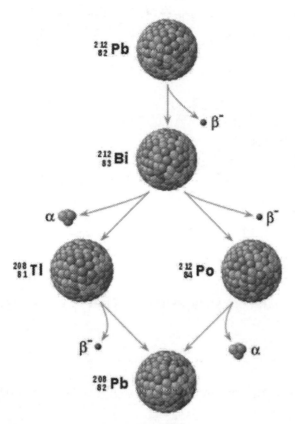

http://en.wikipedia.org/wiki/Radiometric_dating

Radioactive Tracer
>A small amount of radioisotope replacing a nonradioactive isotope of the element in a compound whose path (for example, in the body) or whose decomposition products are to be monitored by detection of radioctivity; also called a radioactive label.

Radioactivity
>The spontaneous disintegration of atomic nuclei.

Reactants
>Substances consumed in a chemical reaction.

Reaction Quotient
>The mass action expression under any set of conditions (not necessarily equlibrium); its magnitude relative to K determines the direction in which the reaction must occur to establish equilibrium.

Reaction Ratio
>The relative amounts of reactants and products involved in a reaction; maybe the ratio of moles, millimoles, or masses.

Reaction Stoichiometry

Description of the quantitative relationships among substances as they participate in chemical reactions.

Reducing Agent

The substance that reduces another substance and is oxidized.

Resonance

The concept in which two or more equivalent dot formulas for the same arrangement of atoms (resonance structures) are necessary to describe the bonding in a molecule or ion.

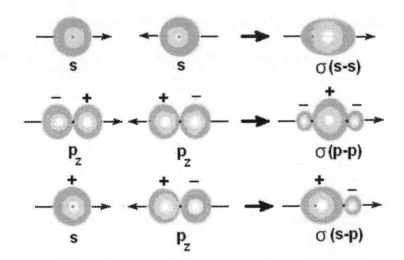

Saturated Hydrocarbons

Hydrocarbons that contain only single bonds. They are also called alkanes or paraffin hydrocarbons.

Saturated Solution

Solution in which no more solute will dissolve.

Semiconductor

A substance that does not conduct electricity at low temperatures but does so at higher temperatures.

Sigma Bonds

Bonds resulting from the head-on overlap of atomic orbitals, in which the region of electron sharing is along and (cylindrically) symmetrical to the imaginary line connecting the bonded atoms.

Sigma Orbital

Molecular orbital resulting from head-on overlap of two atomic orbitals.

Solvent

The dispersing medium of a solution.

S Orbital
A spherically symmetrical atomic orbital; one per energy level.

Specific Gravity
The ratio of the density of a substance to the density of water.

Specific Heat
The amount of heat required to raise the temperature of one gram of substance one degree Celsius.

Spectrum
Display of component wavelengths (colours) of electromagnetic radiation.

Structural Isomers
Compounds that contain the same number of the same kinds of atoms in different geometric arrangements.

Substance
Any kind of matter all specimens of which have the same chemical composition and physical properties.

Substitution Reaction
A reaction in which an atom or a group of atoms is replaced by another atom or group of atoms.

Temperature
A measure of the intensity of heat, i.e. the hotness or coldness of a sample. or object.

Tetrahedral
A term used to describe molecules and polyatomic ions that have one atom in center and four atoms at the corners of a tetrahedron.

Valence Bond Theory
Assumes that covalent bonds are formed when atomic orbitals on different atoms overlap and the electrons are shared.

Chemical bonds formed due to overlap of atomic orbitals

s-s	s-p	s-d	p-p	p-d	d-d
H-H	H-C	H-Pd	C-C	F-S	Fe-Fe
Li-H	H-N		P-P		
	H-F		S-S		

Valence Electrons
Outermost electrons of atoms; usually those involved in bonding.

Valence Shell Electron Pair Repulsion Theory
Assumes that electron pairs are arranged around the central element of a molecule or polyatomic ion so that there is maximum separation (and minimum repulsion) among regions of high electron density.

van der Waals' Equation
An equation of state that extends the ideal gas law to real gases by inclusion of two empirically determined parameters, which are different for different gases.

$$[P + (n^2a/V^2)](V - nb) = nRT$$

Where:

- P - pressure,
- V - volume,
- n - number of moles,
- T - temperature,
- R - ideal gas constant. If the units of P, V, n and T are atm, L, mol and K, respectively, the value of R is 0.0821
- a and b - constants, which are chosen to fit experiment as closely as possible to individual gas molecule. It is similar to PxV =RT

Voltage

Potential difference between two electrodes; a measure of the chemical potential for a redox reaction to occur.

Voltaic Cells

It is an electrochemical cell that derives electrical energy from spontaneous redox reaction taking place within the cell. It generally consists of two different metals connected by a salt bridge, or individual half-cells separated by a porous membrane.

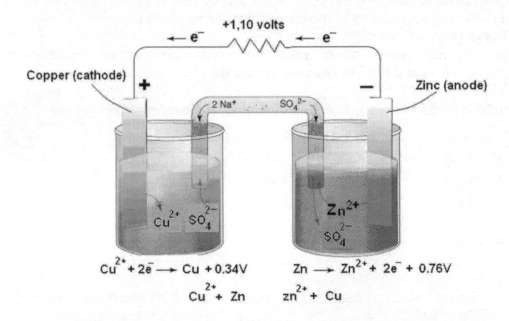

Weak Field Ligand

A Ligand that exerts a weak crystal or ligand field and generally forms high spin complexes with metals.